Sergio Ortolani did his thesis on the dynamics of fast z-pinches at the ENEA Fusion Laboratory in Frascati (Rome), receiving his doctorate in Physics from the University of Rome in 1970. In 1971, he joined the Istituto Gas Ionizzati of the Italian National Council of Research (CNR) in Padova where he began studying plasma magnetic confinement in Reversed Field Pinch configurations as part of the European Fusion Programme. In 1975 he worked at the CTR division of Los Alamos National Laboratory on toroidal pinch experiments. For the last 20 years he has been engaged in experimental and theoretical research on MHD fluctuations and on plasma radiation and transport phenomena, working in close collaboration with many research institutions in Europe, USA, Japan, and USSR. He is currently involved in the study of relaxation phenomena and of heating and transport processes in laboratory and in naturally occurring plasmas. Dr. Ortolani is presently CNR Director of Research at the Istituto Gas Ionizzati (EURATOM-ENEA-CNR Association, Padova) where he heads the scientific programme of the newly-constructed European experiment, RFX.

Dalton D. Schnack received his Ph. D. in Applied Science from the University of California, Davis, in 1977. He did his doctoral research at Lawrence Livermore National Laboratory (LLNL), where he began his interest in non-linear MHD processes in fusion plasmas. From 1973 to 1980 he served as a staff physicist in the computational physics group at the National Magnetic Fusion Energy Computer Center at LLNL. In 1980 he joined the fusion theory group at Los Alamos National Laboratory, where he worked on problems relevant to the reversed field pinch and compact torus experiments. In 1982 he joined Science Applications International Corporation, and in 1986 he was appointed Manager of the Applied Plasma Physics and Technology Division. Dr. Schnack has authored many papers in the fields of linear and non-linear resistive MHD, and computational methods related to such problems. He is actively involved in studying the non-linear MHD properties of the reversed field pinch and the solar corona, in theoretical studies of radiative instabilities in magnetized plasma, and in studying the highly non-linear (turbulent) properties of the Navier–Stokes and MHD equations.

Magnetohydrodynamics of Plasma Relaxation

Sergio Ortolani

Istituto Gas Ionizzati CNR
EURATOM-ENEA-CNR Association
Padova, Italy

Dalton D. Schnack

Science Applications International Corporation
San Diego, California, USA

World Scientific
Singapore • New Jersey • London • Hong Kong

Published by

World Scientific Publishing Co. Pte. Ltd.
5 Toh Tuck Link, Singapore 596224
USA office: 27 Warren Street, Suite 401-402, Hackensack, NJ 07601
UK office: 57 Shelton Street, Covent Garden, London WC2H 9HE

British Library Cataloguing-in-Publication Data
A catalogue record for this book is available from the British Library.

MAGNETOHYDRODYNAMICS OF PLASMA RELAXATION

ISBN-13 978-981-02-0860-8
ISBN-10 981-02-0860-X

Contents

PREFACE ix

1. INTRODUCTION 1

    1.1 Taylor's Theory of Plasma Relaxation 2

    1.2 The Underlying Physics of Plasma Relaxation 6

    1.3 Toroidal Systems and the RFP 9

    1.4 Overview of the Remainder of the Book 12

2. THE RESISTIVE MAGNETOHYDRODYNAMIC MODEL 15

    2.1 Resistive Magnetohydrodynamics 15

        The Physical Model 15

        The Evolution of the Fluid 17

        The Evolution of the Electromagnetic Fields 21

        Characteristic Oscillations; Normal Modes 22

        Dimensionless Variables; the Lundquist Number 25

    2.2 MHD Stability 26

        Linear Stability of Normal Modes 26

        Magnetic Shear and Singular Surfaces 28

        Toroidal Pinch Configurations 29

        Resistive Instabilities 31

        Nonlinear Effects 37

    2.3 Stability Properties of the RFP 39

    2.4 The Force Free MHD Model 42

    2.5 The Role of Numerical Simulation 43

3. TAYLOR'S THEORY OF PLASMA RELAXATION 47

    3.1 The Constraints of Ideal MHD 48

        The Woltjer Constraints 48

        The Topological Properties of the Woltjer Constraints 50

    3.2 Energy Minimization with the Constraints of Ideal MHD 51

3.3   The Effect of Plasma Resistivity... 53

Taylor's Conjecture 53

3.4   Energy Minimization with the Global Helicity Constraint........... 54

Validity of Taylor's Conjecture... 55

Properly Defined Helicity 57

The Role of Plasma Pressure in Taylor's Theory....................... 58

3.5   Predictions of the Theory... 58

The Reversed-Field Pinch... 58

Summary of RFP Predictions.. 61

The Multipinch Experiment 62

3.6   Discussion... 63

4. PHENOMENOLOGY OF RELAXATION IN THE REVERSED-FIELD
PINCH 67

4.1   Mean Field Profiles.. 68

4.2   The Stability of Relaxed States 74

4.3   Resistive Diffusion.. 82

4.4   The Phenomenological Cyclical Model 87

4.5   Experimental Observations of Relaxation Phenomena
in the RFP.. 88

5. THE DYNAMICS OF PLASMA RELAXATION.................................... 95

5.1   Classical Dynamo Theory... 96

Kinematic Dynamos.. 97

Cowling's Theorem... 99

The Turbulent Dynamo 101

Relevance to the RFP Dynamo.. 102

5.2   The Basic Relaxation Mechanism.. 104

The Original Work of Sykes and Wesson................................... 104

Spontaneous and Driven Reconnection in the RFP..................... 106

Fluctuations and Ohm's Law... 110

Evidence of Taylor Relaxation 111

The Helical Ohmic State... 113

5.3   Effects of Nonlinear Mode Coupling................................... 115

MHD Fluctuations... 115

Nonlinear Mode Coupling.. 118

5.4   Summary... 126

6. PRACTICAL ISSUES RELATED TO RELAXATION........................ 129

6.1   Anomalous Loop Voltage.. 130

Perfectly Conducting Outer Boundary............................ 130

Operation with Resistive Walls and Limiters.................. 130

Helicity Balance.. 138

6.2   Taming the Dynamo; An Application of the Theory........ 139

7. RELAXATION AND THERMAL TRANSPORT.............................. 143

7.1   A Model for Sawtooth Oscillations in the RFP............... 144

Experimental Observations.. 144

Theoretical Interpretation of the Sawtooth Crash........... 145

7.2   Thermal Transport During Sawtooth Oscillations.......... 148

Energy Confinement Time.. 148

Modifications to the Resistive MHD Model...................... 149

Simulation of Sawtooth Oscillations............................... 150

7.3   Summary... 154

8. DYNAMICAL RELAXATION IN THE SOLAR CORONA............. 155

8.1   Overview of Coronal Dynamics...................................... 155

8.2   Magnetic Arcade Evolution... 157

8.3   Coronal Current Filaments.. 163

8.4   An Analogy Between the Solar Corona and the RFP....... 170

9. SUMMARY... 171

9.1   Relaxation in the Reversed-field Pinch........................... 172

9.2   Relaxation and Transport.. 174

9.3   Relaxation in the Solar Corona....................................... 175

9.4 Critique.. 175

To what extent can the numerical simulations be believed?....... 176

What about analytic theory? 176

What is the role of turbulence? 176

Are pressure driven modes important?.................................. 177

Are there non-MHD effects? 177

What is the future of relaxation studies?.............................. 177

REFERENCES 179

INDEX 185

Preface

This book describes what the authors have learned about the causes and consequences of plasma relaxation over the past twenty years. Much of this knowledge has come from the field of controlled fusion research, in particular the Reversed-field Pinch (RFP) concept. Thus, a certain bias toward plasma phenomena inherent in this experiment is inevitable. However, it is the authors' belief that these phenomena are sufficiently generic that the concepts presented herein will be useful to researchers in other fields. Failing that, the book is at least an account of a successful world-wide research program that has led to a fairly complete, self-consistent description of a particular type of hot, magnetized plasma.

Early laboratory experiments discovered that axisymmetric toroidal pinch plasma discharges exhibited favorable confinement properties when operated in one of two regimes: the tokamak, in which the toroidal component of the magnetic field was approximately uniform and much larger than the poloidal component; and, the RFP, in which the toroidal field reversed direction in the outer regions of the plasma, and was comparable to the poloidal field. Furthermore, it was found that these states could be formed spontaneously by controlling only a few global parameters, such as the total toroidal current and magnetic flux. This behavior was explained in a general theoretical way by J. B. Taylor in 1974. Taylor used a variational principle with constraints to show that these states were the preferred states of the magnetoplasma system. These states were called relaxed states, and the process of attaining them was called relaxation. Relaxation concepts have since been applied in a variety of situations by a number of researchers to explain the natural states of plasmas and fluids.

The variational theory has proven useful for predicting the time-asymptotic average steady state of many systems, but it does not directly address the fundamental physical processes that give rise to plasma relaxation. Thus, in the ensuing years the fundamental questions shifted from explaining the existence of preferred states to understanding *how* and *why* relaxation occurs. Answering these questions has required a synergism between detailed experimental measurements, large scale numerical computations, and theoretical plasma physics. The results of these efforts are the subject of this book.

Taylor's theory requires only that the dynamical processes responsible for plasma relaxation be describable by the equations of resistive magnetohydrodynamics (MHD). While the specific details of the dynamics do not enter the variational theory, Taylor envisioned plasma relaxation as the result of MHD turbulence.   These fluctuations are characterized by relatively high frequency and short spatial scales.   Subsequent experiments showed that the ubiquitous characteristic fluctuations in RFP discharges have low frequency and broad spatial extent.   These fluctuations were reproduced in large scale computer simulations of RFP discharges.   These simulations also exhibited plasma relaxation.   Theoretical analyses of these results have shown how these long wavelength MHD fluctuations arise, and how they can produce not only the preferred relaxed states predicted by the variational theory, but can also account for many of the operating characteristics of these experiments.   Further calculations have suggested that the dynamics that underlie plasma relaxation may also play an important role in determining the energy confinement properties of RFP plasmas.

The presentation of these results is (we hope) logical rather than historical, and relatively self-contained.   Chapter 1 is devoted to a brief overview of plasma relaxation, toroidal pinches in general, and the RFP in particular.   Chapter 2 presents the fundamentals of the resistive MHD model as required for describing plasma relaxation, including discussions of ideal and resistive instabilities and nonlinear phenomena.   Taylor's variational theory is presented in detail in Chapter 3.   In Chapter 4 we present phenomenological models of plasma relaxation as deduced from experimental observations and linear stability theory.   Typical experimental results are also presented.   The results of detailed, large scale numerical simulations of the magnetic field dynamics in RFP plasmas are presented in Chapter 5.   It is these calculations that self-consistently explain both the presence of the experimentally observed long-wavelength fluctuations and their role in producing plasma relaxation.   A brief discussion of the so-called RFP dynamo is also given here.   The remainder of the book is devoted to the consequences of plasma relaxation, and to relaxation in situations other than fusion experiments.   In Chapter 6 we show that the same fluctuations that are responsible for the attainment of relaxed states can also self-consistently account for anomalous plasma resistance, and enhanced loop voltage when discharges are operated with resistive shells and limiters.   Chapter 7 contains a discussion of the possible consequences that the inherent fluctuations may have for energy transport and confinement.   Chapter 8 describes calculations of similar dynamical events that may occur in the solar corona, the magnetized outer atmosphere of the sun.   While the boundary conditions make it difficult to make a direct connection between these phenomena and the variational relaxation theory, similarities between the dynamics of the coronal and RFP plasmas are noted and discussed.   A brief summary is found in Chapter 9.

Many researchers have contributed to the picture of plasma relaxation that is presented in this book.   To attempt to acknowledge them all would be folly, for some deserving soul would surely be left off any list.   Nonetheless, the authors would like to especially thank their longtime colleagues H. A. B. Bodin, E. J. Caramana, J. Killeen, and R. A. Nebel for providing both insights and a stimulating scientific environment. We wish to acknowledge the specific contributions of our close collaborators V. Antoni, D. C. Barnes, S. Cappello, G. G. Craddock, D. S. Harned, Y. L. Ho, P. Martin, D. Merlin, Z. Mikić, and R. Paccagnella. We would also like to thank S. Prager and C. Gimblett for carefully reading a draft version of the manuscript and providing valuable comments that greatly improved the clarity and accuracy of the presentation. We thank CNR of Italy and the United States Department of Energy for supporting this research over the years. We are indebted to Ms. Anna Miklović for the seemingly endless task of formatting and preparing the camera-ready manuscript. This book exist because of her persistent and professional efforts. One of us (DS) thanks Prof. C. K. Rowdyshrub for spiritual enlightenment. Finally, we emphasize that the interpretation and synthesis of the results presented here is strictly the viewpoint of the authors.

Sergio Ortolani
Padova, Italy

Dalton D. Schnack
San Diego, CA

July 1992

CHAPTER 1
INTRODUCTION

It is observed that many continuous systems naturally evolve toward states that exhibit some form of order on long length scales. Examples are the formation of isolated vortices in two-dimensional Navier-Stokes flow, the appearance of zonal flows in rotating fluids (as observed in the Jovian atmosphere, for example), the evolution of solitons in fluid and optical systems, and the structure of magnetic field configurations in laboratory plasma experiments. In all cases, long range order in one quantity is apparently accompanied by short range disorder in another quantity, so that an overall entropy increase is assured [Hasegawa, 1985]. In many cases these ordered states are remarkably robust in the sense that their detailed structure remains relatively invariant across experimental realizations: the properties of these preferred states are independent of the way the system is prepared. This phenomenon has been called *self-organization* [Hasegawa, 1985].

Self-organization has been most extensively studied in laboratory plasma experiments, particularly within the international program to construct a fusion power reactor. There it has been observed that the magnetic field configurations in a variety of devices tend to evolve toward a small number of preferred states independent of the initial conditions of the system; the detailed structure of the state depends only on a few global parameters, such as total current or magnetic flux, and the inherent geometry of the particular device. These states appear to be reproducible down to the details of the time dependent fluctuations. Examples of such devices are the Reversed-field Pinch (RFP) [Bodin and Newton, 1980], the spheromak [Jarboe et al., 1983], and the tokamak [Furth, 1985]. The occurrence of self-organization in these devices is called *plasma relaxation* [Taylor, 1974]. This book is about the details of plasma relaxation.

It is natural to inquire as to the physical processes that underlie self-organization. Continuous system that exhibit self-organization have several ingredients in common [Hasegawa, 1985]. They are described by nonlinear partial differential equations with dissipation; these equations admit several quadratic or higher order quantities that are conserved in the absence of dissipation; and, these conserved quantities decay at different rates when dissipation is taken into account. In the absence of dissipation the conserved quantities place many constraints on the evolution of this *ideal* system. In the presence of dissipation (or, for time scales on which dissipative effects become important) all but a few (in many cases, one) of the constraints are removed by the selective decay of the ideal invariants, and the system can evolve to a single state determined by the most robust (least dissipated) of the conserved quantities. Theoretically, this final state is often described by a variational principle in which a configuration is found that minimizes one global quantity while another remains invariant.

The variational theory is most developed for the case of plasma relaxation [*Taylor*, 1974, 1986]. The theory is remarkable in that it is able to predict the global state of many experimental plasma configurations with at least qualitative, and in many cases quantitative, accuracy. However, being of a variational nature, it says nothing about the details of the dynamics underlying the relaxation process: it does not tell us what is going on to make the final plasma state come out the way it does. These dynamical details have been the subject of intense study over the past decade. The results of those studies are detailed in later chapters of this book.

1.1 Taylor's Theory of Plasma Relaxation

In this section we briefly introduce Taylor's theory of relaxed plasma states [*Taylor*, 1974, 1986]. Our goal is to provide an overview of the properties of these states, and to motivate the discussion of the underlying dynamics that is to follow. Details of the theory are given in Chapter 3.

In a plasma with no flow, the equations that are appropriate for describing long wavelength, low frequency motions in the absence of dissipation (the equations of *ideal Magnetohydrodynamics*, or MHD) admit an infinite number of conserved quantities. A detailed discussion of these quantities is given in Chapter 3. Here, we simply state the invariants without proof. They are the total energy W

$$ W = \int_{V_0} \left(\frac{B^2}{8\pi} + \frac{p}{\gamma - 1} \right) dV \tag{1.1} $$

and the infinite number of integrals

$$ K_l = \int_{V_l} \mathbf{A} \cdot \mathbf{B} \, dV \,, \quad l = 0, 1, 2 \ldots \tag{1.2} $$

where the integrals are taken over the volume V_l of *each magnetic flux tube* in the plasma [*Woltjer*, 1958]. In Eqs. (1.1-2), $\mathbf{A}$ is the vector potential, $\mathbf{B} = \nabla \times \mathbf{A}$ is the magnetic flux density, p is the pressure, and γ is the ratio of specific heats. The variational problem is then to minimize W with respect to the invariance of the infinite number of integrals given by Eq. (1.2). The details of this calculation are given in Chapter 3. The result is that, for a plasma with negligible internal energy, the relaxed state has magnetic fields that are given by

$$\nabla \times \mathbf{B} = \mu(\mathbf{r})\mathbf{B} \ , \tag{1.3}$$

where

$$\mathbf{B} \cdot \nabla \mu = 0 \ . \tag{1.4}$$

Magnetic fields that satisfy Eq. (1.3) are called *force free*, since the Lorentz force $\mathbf{J} \times \mathbf{B}$ vanishes. Equation (1.4), which follows immediately upon taking the divergence of Eq. (1.3) and using the solenoidal property of the magnetic field, is a statement that μ is constant along flux tubes. Note that μ can be written as a normalized parallel current density $\mathbf{J} \cdot \mathbf{B}/B^2$.

Now, in a fluid with infinite electrical conductivity (zero resistivity, or electrical dissipation) flux tubes retain their integrity for all times. Taylor recognized that when the system is prepared (during the gas breakdown phase of a plasma experiment, for example) there is a value of μ associated with each flux tube in the initial state, and that the *details* of this spatial distribution of μ are uncontrollable; they of necessity vary greatly between different realizations of the experiment. Since the flux tubes retain their integrity in a perfectly conducting fluid, and since μ must remain constant along flux tubes, the function $\mu(\mathbf{r})$ in Eq. (1.3), and hence the final magnetic fields themselves, depend in a detailed manner on the way the system was prepared; the final state is *not* independent of the initial conditions. But this is in contradiction to the observed properties of relaxed states. Clearly, the constancy of the infinite set of integrals Eq. (1.2) unphysically limits the evolution of the system.

Even the smallest amount of dissipation in an electrically conducting fluid allows the flux tubes to merge with each other. Taylor hypothesized that in a real plasma with large but finite electrical conductivity (small but non-vanishing electrical dissipation, or resistivity) the breaking of the flux tubes would render the invariance of the K_l invalid. However, if the plasma is bounded by a perfectly conducting boundary (as is a good approximation in many laboratory experiments), then the only flux tube that retains its integrity is the one tangent to the conducting boundary. Then the *single* invariant

$$K_0 = \int_{V_0} \mathbf{A} \cdot \mathbf{B} \, dV \tag{1.5}$$

would remain, where the integral is now taken over the entire plasma volume. This quantity is called the total *magnetic helicity* of the system. Then minimizing W with respect to this single invariant yields

$$\nabla \times \mathbf{B} = \mu \mathbf{B} \ , \tag{1.6}$$

where μ is now a *constant*. Thus the solutions of Eq. (1.6) are independent of the initial conditions, and can describe physically interesting relaxed states.

In cylindrical geometry, which is a reasonable approximation for many fusion experiments, solutions of Eq. (1.6) that depend only on the radial coordinate are

$$B_r = 0 \ , \tag{1.7a}$$

$$B_\theta = B_0 J_1(\mu r) \ , \tag{1.7b}$$

$$B_z = B_0 J_0(\mu r) \ , \tag{1.7c}$$

where J_0 and J_1 are Bessel functions, and B_0 is a constant. This magnetic field system is called the *Bessel Function Model* (BFM). It can be shown that this solution is in fact a minimum energy state when $\mu a \leq 3.11$, where $r = a$ is the radius of the perfectly conducting boundary. These fields are shown in Figure 1-1.

The BFM can be parameterized in terms of two dimensionless quantities that have proven useful in describing laboratory experiments: the field reversal parameter $F = B_z(a)/\langle B_z \rangle$, and the pinch parameter $\Theta = B_\theta(a)/\langle B_z \rangle$, where $\langle .. \rangle$ represents a volume average. For the BFM we find

$$\Theta = \frac{\mu a}{2} \ , \tag{1.8a}$$

$$F = \frac{\Theta J_0(2\Theta)}{J_1(2\Theta)} \tag{1.8b}$$

The theory thus predicts that when $\Theta > 1.2$ the axial field at the wall will reverse with respect to its value on axis. This phenomenon is observed in the RFP. Another operating regime is given by $F \approx 1$, $\Theta \ll 1$; this is the tokamak. Typical tokamak profiles are also sketched in Figure 1-1. The states for intermediate values of Θ have unfavorable stability properties when a vacuum region lies between the plasma and the conducting boundary, and are thus of less practical interest.

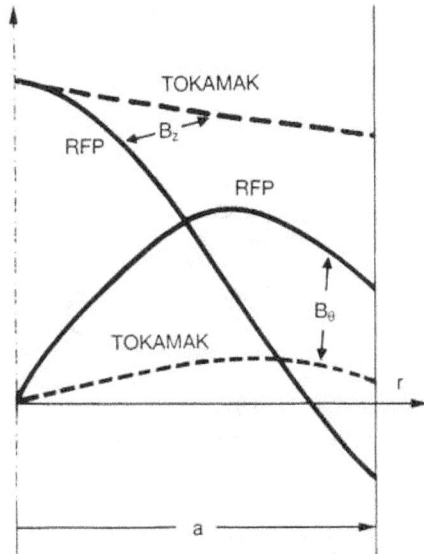

Figure 1-1. Example of the force-free magnetic field profiles for the RFP (BFM with $\mu a = 3$) and the tokamak.

Let us now find the solution of Eq. (1.6) for the case of the tokamak, $F \approx 1$, $\Theta \ll 1$. From Eqs. (1.8), in this regime we have $\mu a \ll 1$, and from Eq. (1.7) we find that $B_z = constant$, $B_\theta = constant \times r$. This state of approximately uniform axial magnetic field and current density is *not* in agreement with experimental results. The inclusion of toroidal effects produces more realistic profiles [*Taylor*, 1984].

The discrepancies between the simple relaxation theory and experiment for the case of the tokamak have been discussed by Kadomtsev [*Kadomtsev*, 1977, 1987]. The theoretical relaxation process as envisioned by Taylor assumes a complete reconnection of flux tubes throughout the entire plasma volume; this is required to argue for the existence of only the single invariant K_0. The resulting completely relaxed states have $\mu = constant$. In practice, such states are never observed due to effects not included in the theory, such as realistic boundary conditions, pressure gradients, and geometry. This is called incomplete relaxation. For example, in the tokamak complete reconnection of flux tubes is precluded by the spatial distribution of the magnetic field. However, we see in Chapter 5 that in the RFP extensive magnetic reconnection may occur quite naturally.

We note that under certain conditions of tokamak operation magnetic fields of the form $B_z = constant$, $B_\theta = constant \times r$ do appear, usually abruptly and with unfavorable consequences for plasma confinement [Kadomtsev, 1987]. This process is called the major disruption.

It may be possible to describe experimentally observed tokamak states using a form of relaxation theory [Kadomtsev, 1987]. The problem is that Taylor's invariant K_0 is not as relevant to the tokamak as it is to the RFP (with the possible exception of the major disruption) because of the incomplete reconnection. Instead, one can minimize the total plasma energy (magnetic plus thermodynamic) with the constraint that the total plasma current remain constant. In that case, certain optimal profiles for pressure and current density are found [Kadomtsev, 1987]. This *profile consistency* has in fact been observed in many experimental tokamak plasmas [Murakami et al., 1985; Bickerton et al., 1986]. Other phenomena, such as the so-called L-H transition [Kailhacker et al., 1985], may also be describable in terms of plasma relaxation [Kadomtsev, 1987].

1.2 The Underlying Physics of Plasma Relaxation

We have seen that self-organized, or relaxed, states of continuous physical systems can be found by minimizing a particular integral, usually related to the energy, with the constraint that another (or at most a few) global integral remains fixed. The choice of the proper constraining condition clearly is determined by some underlying physics. The constraints that work for one system may not work for another. In fact, we have seen that Taylor's global helicity invariant K_0, which works well in the RFP, appears to have less relevance to normal tokamak operation. Nonetheless, it is of interest to attempt to find some unifying thread in the physics that governs the relaxation process in so many situations.

One point of view is that natural long length scale self-organization is generated from the turbulent fluctuations of the system. This turbulence is random and short scale, and is characterized by its statistical properties. The equations that govern the motion of the medium, and hence also its turbulent properties, contain several quantities of quadratic or higher order that are conserved in the absence of dissipation. When this occurs, it is possible for one of the conserved quantities, say A, to accumulate at (or cascade to) short spatial scales, and the other, say B, to cascade to long spatial scales [Kraichnan, 1967; Montgomery et al., 1978]. In the presence of dissipation that occurs at short spatial scales, such as that related to viscosity or resistivity, A is effectively destroyed, while B remains relatively invariant. This process is called *selective decay*; the system preserves one global integral at the expense of others. Then for such a system we would expect A to be minimized while B remains relatively constant, leading to a variational principle of the type described in Section 1.1. In this sense self-organization is relative. The long range order in the system applies to B alone; the system is completely disordered when viewed from the perspective of A [*Hasegawa*, 1985].

For example, in *two-dimensional* Navier-Stokes flow, both the kinetic energy, $W = 1/2 \int \rho v^2 dV$, and the mean square vorticity, or enstrophy, $U = 1/2 \int (\nabla \times v)^2 dV$, are conserved in the absence of viscous dissipation. In the presence of viscosity, U cascades to short wavelengths and is dissipated while W cascades to long wavelengths and remains relatively invariant. Thus, the appropriate variational principle is $\delta U - \lambda \delta W = 0$. This leads to the equation

$$\nabla \times \nabla \times v = \lambda v , \qquad (1.9)$$

which is analogous to Eq. (1.6). The accumulation of energy in large eddies has been born out by solutions of Eq. (1.9), as well as by detailed computer simulations of the time-dependent equations [*Fornberg*, 1977; *Hossain et al.*, 1983].

It is interesting to note that inverse cascades have not been observed in *three-dimensional* Navier-Stokes flow. This may be related to the fact that the relevant conserved quantity in this case is the *total helicity*, $H = 1/2 \int v \cdot \nabla \times v dV$, which is not positive definite [*Hasegawa*, 1985]. The nonexistence of inverse cascades for three-dimensional flows has also been argued from statistical grounds [*Kraichnan*, 1973].

The large scale dynamics of many experimental plasma systems are well described by the equations of MHD [*Freidberg*, 1987]. For three-dimensional resistive MHD, the proper invariants are the magnetic energy, Eq. (1.1), and the magnetic helicity, Eq. (1.5). For such a system a turbulent inverse cascade of magnetic helicity to long wavelength and magnetic energy to short wavelength has been demonstrated [*Frisch et al.*, 1975; *Pouquet et al.*, 1976; *Montgomery et al.*, 1978]. This leads naturally to a variational principle of the type proposed by *Taylor* [1974, 1986], and would seem to describe the physical process underlying plasma relaxation.

If small scale turbulence and inverse cascades *are* the cause of plasma relaxation, then we should expect that the variational theory should apply equally well to *all* systems that are describable by resistive MHD. This is because the breaking of flux tubes caused by the turbulence would occur at such small scales that the process would not be affected by the specific geometry in which it evolves.

However, we have seen in Section 1.1 that the theory does not apply well to tokamak operation. This may be because the fundamental motions that cause the magnetic reconnection operate differently in the RFP and the tokamak [*Kadomtsev*, 1977]. The dominant magnetic fluctuations observed in both RFPs and tokamaks are long wavelength MHD modes [*Watt and Nebel*, 1983; *Furth*, 1985]. These motions therefore are affected by the geometry of the different devices. The primary relaxation mechanism may therefore be provided by long wavelength, low frequency, deterministic MHD fluctuations, as opposed to small scale statistical turbulence. Throughout this book we adopt this point of view; namely, that long wavelength MHD modes are the principle cause of plasma relaxation. Our goal is to demonstrate the extent to which plasma relaxation in specific geometries can occur as a result of these motions.

One can envision how relaxation might occur as a result of these long wavelength unstable MHD modes. Consider an RFP plasma in a relaxed state. At every point the gradient of the current density is equal to a critical value $dJ/dr = (dJ/dr)_c$; deviations from this value result in unstable MHD modes that drive the current gradient back toward its critical value. The evolution away from the critical profile is naturally provided by resistive diffusion. Similarly, in a tokamak discharge the pressure profile is determined by some critical profile $dp/dr = (dp/dr)_c$; deviations now drive modes that restore the profile [*Kadomtsev*, 1987]. Thus real plasmas exist in a state of constant tension between effects of diffusion driving them away from relaxed states and instabilities restoring them. Indeed, sawtooth-like oscillations are characteristic of many laboratory plasmas.

Over the past decade considerable effort has been expended to elucidate the detailed role of these long wavelength MHD modes in plasma relaxation. This effort has been concentrated on the RFP where Taylor's theory seems most applicable. As a result, a self-consistent picture of nonlinear, long wavelength MHD motions in this device has become mature.

A goal of this book is to demonstrate how these long wavelength, low frequency, coherent MHD fluctuations can produce plasma relaxation. We shall see that these motions are the nonlinear state of linearly unstable normal modes of the system. Through their nonlinear dynamical interaction they produce the magnetic field configurations and fluctuations that are observed experimentally. The inherent nonlinearity of the process renders analytic treatment virtually impossible. As a result, numerical simulation plays a central role in determining the picture that will be presented. In fact, our present understanding of this relaxation process relies to a great extent on advanced numerical algorithms and supercomputers.

The difficulties inherent in this type of numerical simulation are briefly discussed in Chapter 2. For now, however, we simply state that *fully turbulent numerical MHD simulations of naturally occurring and laboratory plasmas are beyond the capability of present computing technology.* However, long wavelength motions can be accurately resolved. In these calculations, the inevitable buildup of energy at the small length scales due to quadratic nonlinearities is controlled by some form of artificial viscosity.

Based on this type of calculation, we attempt to demonstrate that plasma relaxation in certain configurations is the result of long wavelength modes. However, it can be argued that turbulent relaxation cannot be ruled out because, in fact, the numerical simulations on which these conclusions are based cannot assess the effects of the turbulence. This is indeed true. It is also true that numerical simulations of long wavelength modes demonstrate plasma relaxation, and reproduce at least qualitatively most experimentally measured characteristics. It can thus be said that relaxation in *numerical* plasmas can arise from long wavelength motions. The future will judge the extent to which this result fully applies to *real* plasmas.

1.3 Toroidal Systems and the RFP

In Section 1.1 we briefly introduced Taylor's relaxation theory, and suggested that it may provide a basis for describing the RFP. The relaxation process has been most systematically studied in this device. The extent to which the theory agrees with experimental results is discussed in detail in Chapter 4. Therefore, in this book we consider the RFP as a paradigm for describing the relaxation process, and concentrate our attention there. It is our hope that the successes (or failures) of describing plasma relaxation in this geometry will provide some insights into the fundamental dynamics responsible for relaxation, and perhaps serve as a guide for detailed studies of relaxation in other systems. In this section we give a brief overview of the characteristics of the RFP plasma.

The goal of magnetic fusion research is to confine a hot plasma with a magnetic field for a sufficiently long time to allow enough fusion reactions to occur so that the resulting energy output exceeds the energy required to operate the confining apparatus. In the most studied confinement concepts, the plasma is in the shape of an axisymmetric torus. Generically, these are called toroidal pinches. Two specific types of toroidal pinches have been studied in detail: the *tokamak* [*Furth*, 1985], and the *RFP* [*Bodin and Newton*, 1980]. They differ in the details of the magnetic field internal to the toroidal plasma. A sketch of a toroidal plasma configuration is shown in Figure 1-2.

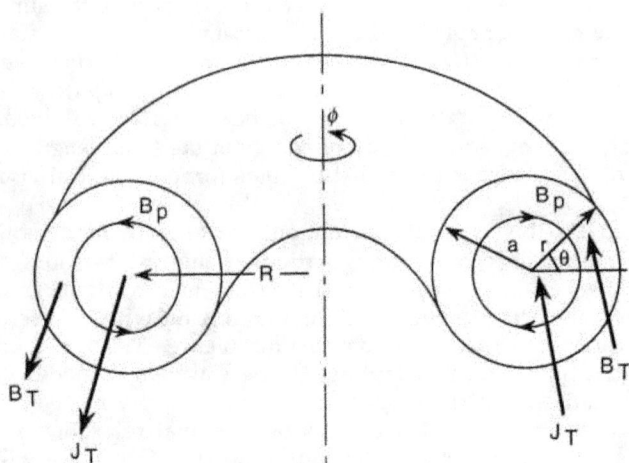

Figure 1-2. Toroidal plasma configuration.

The torus is described by its major radius R, the distance from the axis of rotation (the major axis) to the center of the plasma (the minor axis), and the minor radius a, the distance from the center of the plasma to its outer boundary. The ratio R/a is called the aspect ratio. A convenient coordinate system consists of the minor radial coordinate r, which ranges from the center of the plasma ($r = 0$) to the outer boundary $r = a$, the poloidal angle θ, which measures rotations about the minor axis, and the toroidal angle ϕ, which measures rotations about the major axis. Both θ and ϕ vary between 0 and 2π. The complete toroidal distance along the minor axis (corresponding to one rotation about the major axis) is $2\pi R$. Clearly, all physical quantities must be periodic in θ and ϕ.

Equilibria in toroidal plasmas that are independent of the toroidal angle ϕ are called axisymmetric. In Chapter 2 we will see that for axisymmetric equilibria the magnetic field lines form a set of nested toroidal surfaces. These surfaces are called *flux surfaces*. In general, these nested tori do not share a common geometric center. Furthermore, none of these geometric centers correspond with the minor axis $r = 0$; the surfaces are shifted toward the outer edge of the torus. The amount of this shift is dependent on the inverse aspect ratio $\varepsilon = a/R$, on the relative amount of plasma confined by the toroidal current, as measured by the poloidal beta parameter $\beta_p = 8\pi p/B_\theta^2$, and on the current density distribution. These equilibria are functions of the two coordinates r and θ.

In the tokamak, equilibrium (radial force balance) is achieved by a balance between the radial pressure gradient and the stresses produced by the poloidal magnetic field. Thus the poloidal beta $\beta_p = 8\pi p / B_\theta^2$ in a tokamak is of order unity. It turns out that a large toroidal magnetic field must then be applied to stabilize the $m = 1$ ideal kink mode. (The theory of MHD stability is briefly discussed in Chapter 2.) This field must be large enough to assure that the safety factor $q \approx rB_\phi / RB_\theta$ is greater than unity everywhere in the device. Thus $B_\phi / B_\theta \gg 1$, and the total beta is $\beta = 8\pi p / B_\phi^2 \approx (r/R)^2 \ll 1$. In the RFP, equilibrium results from a balance between the stresses in the poloidal and toroidal magnetic fields; then $B_\phi / B_\theta \approx 1$, and $q \approx r/R \ll 1$. Finite plasma pressure arises from any small imbalance in these stresses. Therefore, to a good approximation, the RFP is characterized by nearly force-free magnetic fields ($\mathbf{J} \times \mathbf{B} \approx 0$), and $\beta_p \approx \beta \approx r/R$. Linear stability against ideal kink modes is achieved by a combination of high magnetic shear $((dq/dr)/q)$ and the presence of a close fitting conducting outer boundary that effectively inhibits plasma displacements. (This is described in more detail in Chapters 2 and 4.)

The dependence of toroidal equilibria on two coordinates, as discussed previously, makes the analysis of such systems relatively difficult. However, if the aspect ratio is very large, or if β_p is small, then the outward toroidal shift described previously is small. In that case it is a good approximation to replace the torus with a straight, periodic cylinder; r and θ retain their meanings as minor radius and poloidal angle, while the toroidal coordinate ϕ is replaced by the axial coordinate z that varies between 0 and $2\pi R$. The relevant equations and ensuing analysis are then greatly simplified. Again, all physical quantities must now be periodic in both θ and z.

In present toroidal plasma experiments, the aspect ratio is in the range of approximately three to five. This alone is generally insufficient to justify the cylindrical approximation. However, in the RFP the toroidal magnetic field is relatively weak and $q \ll 1$. In this case the poloidal asymmetry in $|\mathbf{B}|$ is much weaker than in a tokamak, and the cylindrical approximation is justified even for relatively small aspect ratio. We use this approximation throughout the remainder of this book; we do so even for the tokamak, since many important concepts relevant to plasma relaxation can be discussed without reference to toroidal effects.

In the RFP it is observed that, with sufficient applied toroidal voltage, the field-reversed state is achieved naturally and spontaneously without the necessity of externally applied poloidal voltages. Once field reversal is achieved we might expect the lifetime of the profiles to be determined by resistive diffusion. However, it is an experimental fact that such plasmas can be maintained for relatively long times at fixed values of F and Θ. The lifetime of the discharge is limited primarily by the volt-seconds available in the external circuit, and plasma density control. The applied electric field supplies poloidal flux to the plasma, which is spontaneously and dynamically converted into axial flux in sufficient amounts to maintain F and Θ at their proper values. This process is called the *RFP dynamo*, because the anomalously long magnetic field lifetime observed in this device is reminiscent of the magnetic field generation that occurs in the earth, and in stars: the classical, or astrophysical, dynamo. Over the past several years, a theoretical understanding of the RFP dynamo has emerged. As a result, it is now believed that this dynamo is fundamentally a cyclic plasma relaxation process, and has significant differences from the classical dynamo. These issues are discussed more thoroughly in Chapters 3, 4, and 5.

1.4 Overview of the Remainder of the Book

In the following chapters we present the dynamical description of plasma relaxation that has arisen from detailed studies of the RFP. Because of its importance in describing relaxation, the resistive MHD model is briefly developed in Chapter 2. The fundamental equations are derived, and the importance of non-vanishing electrical resistivity is emphasized. This allows us to briefly develop a heuristic description of resistive instabilities, the modes that are primarily responsible for the flux tube reconnection that is at the heart of plasma relaxation. We also introduce some fundamental descriptive nomenclature.

In Chapter 3 we present the details of Taylor's variational theory of plasma relaxation. The details of the variational calculation are given, as are further results relevant to both the RFP and other experimental devices. The success and failures of Taylor's theory with respect to the RFP are also discussed.

In Chapter 4 we discuss various phenomenological models of RFP relaxation that have been developed to interpret experimental data. This allows us to further develop the characteristics of RFP operation, and to describe specific experimental results. While these models have been developed solely on the basis of simple linear stability theory and experimental observations, they predict many of the detailed dynamical results that have been obtained with large scale computer simulation.

In Chapter 5 we present the dynamical theory of plasma relaxation in the RFP. The results presented in this chapter have arisen from extensive large scale computer simulations of three-dimensional MHD in RFP geometry. These results address the dynamics of the magnetic and velocity fields. Here we show in detail how the various unstable normal modes nonlinearly interact to produce the richly dynamical state observed in RFP experiments. We show how these modes reproduce the general predictions of Taylor's theory, as well as how they compare with the theory in detail.

In Chapter 6 we present further details of the relaxation dynamics as they relate to practical issues. In particular, we show that the magnetic fluctuations that produce plasma relaxation can also produce anomalously large loop voltage requirements in laboratory experiments. We further address issues relating to experimental operation with a resistive outer boundary. We also describe a conceptual procedure based on the predictions of the dynamical theory that could be used to control fluctuations and drive current in these devices.

In Chapter 7 we discuss the possible connection between the natural magnetic fluctuations necessary for plasma relaxation and the transport of thermal energy through the plasma. Sawtooth oscillations as they occur in the RFP are discussed. We show that plasma relaxation can cause anomalously large thermal transport, even when the individual transport coefficients have classical values.

As an example of a naturally occurring plasma, in Chapter 8 we discuss relaxation in the solar corona. Calculations relevant to the dynamics of both active and quiet coronal regions are presented, and an analogy between the RFP and the corona is given.

Finally, in Chapter 9 we give a summary of our understanding of the dynamics of plasma relaxation, and discuss various issues that are as yet not well understood.

CHAPTER 2

THE RESISTIVE
MAGNETOHYDRODYNAMIC MODEL

In the previous chapter, the concept of plasma relaxation was introduced. In this chapter, we begin a theoretical description of these relaxed plasma states. While general conclusions can be drawn about these states, an understanding of the dynamical details of the relaxation process has required that both experiments and large scale numerical simulation techniques be concentrated on specific magnetoplasma configurations. We therefore begin in Section 2.3 to focus most of our attention on the RFP plasma; the theory is most applicable to this particular device. This continues into the following chapters, where the detailed dynamics of plasma relaxation in the RFP are presented.

We begin this chapter with a derivation of the equations of resistive MHD. From the outset, we treat the plasma as a continuous, conducting fluid that satisfies the classical laws of motion and thermodynamics. The electromagnetic field is described by neglecting the displacement current in Maxwell's equation, as is a reasonable assumption for low frequency phenomena. Dimensionless variables are introduced, and the significance of the Lundquist number is described. This discussion is relatively self-contained. This is followed by a brief review of ideal and resistive MHD stability theory. The emphasis here is on physical principles, so the discussion is heuristic rather than rigorous. A short review of linear stability results relevant to the RFP is also presented. The concepts introduced here play a major role in the description of relaxation dynamics given in later chapters. For a more detailed discussion of the MHD model, the reader is referred to the books by *Bateman* [1978] and *Freidberg* [1987] and the references cited therein.

2.1 Resistive Magnetohydrodynamics

The Physical Model

As described in Chapter 1, the dominant magnetic fluctuations that characterize plasma relaxation are apparently low frequency and spatially global. The simplest self-consistent model describing the macroscopic behavior of a plasma is *magnetohydrodynamics* (MHD). This model combines fluid equations, similar to those used in fluid dynamics, with a

15

form of Maxwell's equations. In this model, the state of the magnetoplasma system is described by specifying the mass density ρ, the velocity $\mathbf{v}$, the magnetic field $\mathbf{B}$, and either the scalar pressure p or the internal energy density e, as functions of space and time.

In the simplest MHD model, transport processes, such as those arising from finite plasma resistivity, are neglected. The role of this *ideal* MHD model is to investigate the equilibrium and stability properties of various configurations. Nonideal effects, such as finite resistivity, thermal conduction, and viscosity, may allow instabilities to develop that are slower and weaker than those allowed by ideal MHD. In magnetic fusion, these processes can lead to either enhanced transport or violent premature termination of an experimental plasma discharge. These also play a role in astrophysical and space plasmas. Examples are the sudden, violent release of magnetic energy from the solar corona resulting in a solar flare, and magnetic substorms in the earth's magnetosphere. Nonideal effects due to finite plasma resistivity are also essential to the dynamics of plasma relaxation. We are therefore interested in developing a model that includes these effects.

The MHD model, in most cases, provides the simplest meaningful description of macroscopic plasma behavior for time scales shorter than those determined strictly by diffusive transport. Despite its apparent simplicity, the model is rich in dynamical behavior and subtleties.

The MHD model is derived by considering the plasma to be an electrically conducting, charge neutral fluid moving in response to electromagnetic and pressure forces, as described by Newtonian mechanics. Such a description implicitly assumes that the fluid can be subdivided into infinitesimally small fluid elements, each of which has the same material properties as the macroscopic system. The macroscopic equations governing the flow are then determined by applying the appropriate laws of motion and thermodynamics to each of these fluid elements. This assumption breaks down if the size of the fluid elements becomes smaller than the Debye length λ_D, the length over which nonuniformities in charge density are expected to be noticeable. Similar conclusions hold for length scales that describe the motion of individual particles, such as the ion gyro-radius ρ_i. Thus, if Δ is the size of an infinitesimal fluid element, the fluid description is valid for length scales L such that $L \gg \Delta \gg \rho_i, \lambda_D$. Furthermore, we assume that conditions are such that relativistic and quantum effects can be ignored. Our goal is to derive a closed set of partial differential equations that describe the evolution of the quantities ρ, $\mathbf{v}$, $\mathbf{B}$, and p.

The Evolution of the Fluid

Consider a fixed macroscopic volume V_0, with bounding surface S_0, of fluid with mass density ρ. The total mass within V_0 is

$$M_0 = \int_{V_0} \rho dV .$$
(2.1)

The rate of mass flow through an element dS of S_0 is $\rho \mathbf{v} \cdot d\mathbf{S}$, and the total mass flowing out of V_0 per unit time is

$$-\frac{\partial M_0}{\partial t} = \oint_{S_0} \rho \mathbf{v} \cdot d\mathbf{S} .$$
(2.2)

Using Eq. (2.1) and Green's Theorem, we find

$$\int_{V_0} \left[\frac{\partial \rho}{\partial t} + \nabla \cdot (\rho \mathbf{v}) \right] dV = 0 .$$
(2.3)

Since Eq. (2.3) must hold for an arbitrary volume V_0, the integrand must vanish identically. We thus find that the law of conservation of mass can be expressed in terms of the partial differential equation

$$\frac{\partial \rho}{\partial t} + \nabla \cdot (\rho \mathbf{v}) = 0 ,$$
(2.4)

which is often called the *continuity equation*.

We now seek a similar equation that describes the motion of an electrically conducting fluid in response to applied forces. We imagine that no mass enters or leaves a fluid element as it moves about in space (the element is said to be *comoving* with the fluid). The equation of motion then simply equates the rate of change of momentum per unit volume of the fluid element, $\rho d\mathbf{v}/dt$, to the applied forces per unit volume.

We shall consider two types of forces acting on the fluid element. The first type is the result of the interaction of the fluid element with its neighbors. These are generally given in units of force per unit area. These forces are expressible in terms of pressure forces acting normal to the surface of the fluid element, and stress forces acting tangential to the surface of the fluid element. (The latter may result from internal friction, or viscosity.) The second type of force acts throughout the volume of the fluid element, and is generally caused by external agencies. These are generally expressed in units of force per unit volume, and are called body forces. Examples of such forces are gravity (ρg), and the Lorentz force of electromagnetism (in cgs units, this force is $J \times B/c$, where J is the current density, B is the magnetic flux density, and c is the speed of light).

The net force exerted on the surface of the fluid element by the pressure and stress forces is

$$F_s = -\oint_{S_0} (p\mathbf{I} - \Pi) \cdot d\mathbf{S} = -\int_{V_0} \nabla \cdot (p\mathbf{I} - \Pi) \, dV \, , \tag{2.5}$$

where I is the unit tensor, p is the scalar pressure, and Π is the stress tensor (a measure of the tangential forces of internal friction; for more details about Π, the reader is referred to the excellent book by *Landau and Lifshitz* [1959]). These surface stresses thus exert an equivalent body force of $- \nabla \cdot (p\mathbf{I} - \Pi)$ per unit volume. The net force per unit volume acting on the fluid element is then the sum of this equivalent body force and the applied body forces. The equation of motion for the fluid element can thus be written as

$$\rho \frac{d\mathbf{v}}{dt} = -\nabla p + \nabla \cdot \Pi + \frac{1}{c} \mathbf{J} \times \mathbf{B} + \rho \mathbf{g} \, . \tag{2.6}$$

Equation (2.6) is the equation of motion expressed in a frame of reference attached to, and moving with, the fluid element. The time derivative $d\mathbf{v}/dt$ on the left-hand-side is the acceleration as measured by an observer who is comoving with the fluid. It does not represent the acceleration at any fixed point in space. This dynamical description is called *Lagrangian*. For our applications, we are interested in relating $d\mathbf{v}/dt$ to the acceleration as measured at a fixed point in space by an observer at rest in the reference frame of the laboratory. This dynamical description is called *Eulerian*, and the corresponding acceleration is denoted $\partial\mathbf{v}/\partial t$.

Physically, the change in the velocity of a fluid element $d\mathbf{v}$ that occurs in a time dt is composed of two parts [*Landau and Lifshitz*, 1959]. The first is the change during dt of the velocity at a fixed point in space:

$$d\mathbf{v}_1 = \frac{\partial\mathbf{v}}{\partial t} dt \, . \tag{2.7a}$$

The second is the difference between the velocities at the same instant of time a distance $d\mathbf{r}$ apart:

$$d\mathbf{v}_2 = d\mathbf{r} \cdot \nabla \mathbf{v} \, . \tag{2.7b}$$

Thus,

$$\frac{d\mathbf{v}}{dt} = \frac{d\mathbf{v}_1}{dt} + \frac{d\mathbf{v}_2}{dt}$$

$$= \frac{\partial \mathbf{v}}{\partial t} + \frac{d\mathbf{r}}{dt} \cdot \nabla \mathbf{v}$$

$$= \frac{\partial \mathbf{v}}{\partial t} + \mathbf{v} \cdot \nabla \mathbf{v} \ . \tag{2.8}$$

Alternatively, Eq. (2.8) can be derived by writing $\mathbf{v} = \mathbf{v}[\mathbf{r}(t),t]$ (where $\mathbf{r}(t)$ is the location of the fluid element as it moves about) and applying the chain rule to differentiation with respect to time. The Eulerian equation of motion is thus

$$\rho \left(\frac{\partial \mathbf{v}}{\partial t} + \mathbf{v} \cdot \nabla \mathbf{v} \right) = -\nabla p + \nabla \cdot \Pi + \frac{1}{c} \mathbf{J} \times \mathbf{B} + \rho \mathbf{g} \ . \tag{2.9}$$

Expressions of the form Eq. (2.8) relating time derivatives as measured in the frame moving with the fluid element to those measured in the lab frame appear throughout fluid mechanics and MHD. For example, by expanding the divergence operator the Eulerian continuity Eq. (2.4) can be written as

$$\frac{d\rho}{dt} \equiv \frac{\partial \rho}{\partial t} + \mathbf{v} \cdot \nabla \rho = -\rho \nabla \cdot \mathbf{v} \ , \tag{2.10}$$

which is the Lagrangian form of the law of conservation of mass. Equation (2.10) states that the mass density of a fluid element (with fixed total mass) can only change as a result of changes in its volume (compression or dilation), as expressed by the right-hand-side, since mass cannot flow through its surface as it moves about. (Flows for which $\nabla \cdot \mathbf{v} = 0$ are thus called *incompressible*.)

Ignoring for the moment $\mathbf{J}$, $\mathbf{B}$, Π, and $\mathbf{g}$, the state of the fluid is specified by the five quantities ρ, $\mathbf{v}$ (three components), and p. (We shall return to $\mathbf{J}$, $\mathbf{B}$, Π, and $\mathbf{g}$ shortly.) Equations (2.4) and the three components of Eq. (2.9) are four equations that describe the evolution of these state variables in space and time. In order to close the system of equations (so that we have as many equations as unknowns) we need to include the laws of thermodynamics as they relate to the fluid element. As with the equation of motion, these laws are most easily expressed in the Lagrangian representation.

An appropriate equation can be obtained by equating the rate of heat input to the fluid element to the rate at which energy flows across its surface (written as $-\nabla \cdot \mathbf{q}$, where $\mathbf{q}$ is the heat flux vector), plus the power of various sources and sinks of energy. Thus

$$\rho \frac{dQ}{dt} = -\nabla \cdot \mathbf{q} + \Pi : \nabla \mathbf{v} + \eta J^2 , \qquad (2.11)$$

where Q is heat per unit mass, and the last two terms on the right-hand-side represent the rates of viscous and Ohmic heating, respectively. Here we have used the notation $\Pi : \nabla \mathbf{v} = \Pi_{ik} \, \partial v_i / \partial x_k$.

The incremental change in heat per unit mass, dQ, can be written as the sum of the work per unit mass done on the fluid element and the incremental change in internal energy per unit mass of the fluid element:

$$dQ = pd\left(\frac{1}{\rho}\right) + de . \qquad (2.12)$$

Using Eqs. (2.10-2.12), we find

$$\rho \frac{de}{dt} = -p\nabla \cdot \mathbf{v} - \nabla \cdot \mathbf{q} + \Pi : \nabla \mathbf{v} + \eta J^2 , \qquad (2.13)$$

which expresses the rate of change of internal energy per unit volume of the fluid element in terms of the work required to compress or dilate the volume, the rate of heat flow across the surface, and the sources and sinks of energy.

We can use the *ansatz* Eq. (2.10) to express Eq. (2.13) in Eulerian form. The result is

$$\frac{\partial}{\partial t} \rho e = -\nabla \cdot (\rho e \mathbf{v}) - p\nabla \cdot \mathbf{v} - \nabla \cdot \mathbf{q} + \Pi : \nabla \mathbf{v} + \eta J^2 . \qquad (2.14)$$

We have now succeeded in obtaining an additional equation, Eq. (2.14), describing the evolution of the fluid, but we have also added an additional unknown: the specific internal energy density e. We must use an *equation of state* to express a further relationship satisfied by the state variables $\rho, \mathbf{v}, p,$ and e. This relationship depends on the material properties of the fluid under study. In studies of relaxation, we are interested in plasma dynamics. To an excellent approximation, a plasma behaves like a perfect gas: a gas composed of non-interacting point particles, each with zero volume [*Krall and Trivelpiece*, 1973]. In that case, $p, \rho,$ and e satisfy the relationship

$$\rho e = \frac{p}{\gamma - 1} , \tag{2.15}$$

where γ is the ratio of the specific heat at constant pressure to the specific heat at constant volume. For a plasma, $\gamma = 5/3$.

Furthermore, for many purposes the effects of thermal conduction, as given by $\mathbf{q}$, and viscosity or other sources of anisotropic pressure Π can be ignored. While this assumption is not strictly valid from physical arguments, the inclusion of these effects in the model does not introduce any fundamentally new dynamical processes and results in an unnecessarily complicated set of equations. Thus for much of our discussion we set $\Pi = \mathbf{q} = 0$. The effects of non-zero heat transport are discussed in Chapter 7. It should be emphasized that plasma resistivity is *not* neglected, even though it may be quite small. This is because even the smallest amount of resistivity fundamentally changes the character of the equations and leads to new, important solutions that are topologically forbidden in a perfectly conducting plasma. The role of resistivity in the dynamics is further discussed later in this chapter. (The resistivity η can be taken to be a specified function of the dependent variables, or can simply be assumed constant.) In laboratory applications gravity can also be ignored. Under these assumptions, the energy equation takes the form

$$\frac{\partial p}{\partial t} = -\gamma p \nabla \cdot \mathbf{v} - \mathbf{v} \cdot \nabla p + (\gamma - 1) \eta J^2 . \tag{2.16}$$

Note that when $\eta = 0$, Eqs. (2.4) and (2.16) imply that the *adiabatic law*, $p/\rho^\gamma = constant$, holds for each fluid element.

The Evolution of the Electromagnetic Fields

Equations (2.4), (2.9), and (2.16) describe the evolution of the fluid state variables ρ, $\mathbf{v}$, and p in response to the Lorentz force. It remains to self-consistently describe the evolution of the electromagnetic fields $\mathbf{B}$ and $\mathbf{E}$, and the current density $\mathbf{J}$, in response to the fluid motion. The evolution of the electromagnetic fields is described by Maxwell's equations. In the MHD model these equations must be slightly modified.

Recall that we have assumed that all motions are non-relativistic. Under this assumption, the equations describing the fluid motion are Galilean invariant. However, Maxwell's equations describe relativistic phenomena, so that they are Lorentz invariant. In order to obtain a consistent set of equations describing the evolution of the coupled system, we must either use a relativistic description of the fluid motion, or a non-relativistic description of the electromagnetic field.   Since the characteristic velocities are much smaller than the speed of light, we choose the latter and ignore the displacement current in Ampére's law.   Thus the propagation of light waves is not described by the model.   The resulting "pre-Maxwell" equations (so called because they represent the state of knowledge of electromagnetism prior to the work of Maxwell) are:

$$\frac{1}{c}\frac{\partial \mathbf{B}}{\partial t} = -\nabla \times \mathbf{E} ,$$
(2.17)

$$\frac{4\pi}{c}\mathbf{J} = \nabla \times \mathbf{B} ,$$
(2.18)

and

$$\nabla \cdot \mathbf{B} = 0 .$$
(2.19)

Note that since Eq. (2.17) implies that $\nabla \cdot \mathbf{B}$ does not change in time, Eq. (2.19) should be regarded as an initial condition and not as an independent equation.

The simplified electron equation of motion (Ohm's Law),

$$\mathbf{E} + \frac{1}{c}\mathbf{v} \times \mathbf{B} = \eta \mathbf{J} ,$$
(2.20)

now serves to couple the evolution of the fluid and the fields.

Equations (2.4), (2.9), (2.16) through (2.18), and (2.20) supplemented by the initial condition Eq. (2.19), constitute a set of fourteen partial differential equations describing the evolution of the fourteen quantities $\rho, p, \mathbf{v}, \mathbf{B}, \mathbf{E}$, and $\mathbf{J}$ in space and time, and are the basic equations of resistive MHD.

Characteristic Oscillations; Normal Modes

Now let us consider the characteristic oscillations supported by the MHD model. For now we ignore the effects of resistivity; its influence on the stability of the normal modes will be discussed in detail later in this chapter.

With $\eta = 0$, the equations of the MHD model can be combined to yield a system of eight equations in the eight unknowns $\rho, \mathbf{v}, \mathbf{B}$, and p:

$$\frac{\partial \rho}{\partial t} + \nabla \cdot (\rho \mathbf{v}) = 0 \ , \tag{2.21a}$$

$$\rho \left(\frac{\partial \mathbf{v}}{\partial t} + \mathbf{v} \cdot \nabla \mathbf{v} \right) = -\nabla p + \frac{1}{4\pi} (\nabla \times \mathbf{B}) \times \mathbf{B} \ , \tag{2.21b}$$

$$\frac{\partial p}{\partial t} = -\gamma p \nabla \cdot \mathbf{v} - \mathbf{v} \cdot \nabla p \ , \tag{2.21c}$$

$$\frac{\partial \mathbf{B}}{\partial t} = \nabla \times (\mathbf{v} \times \mathbf{B}) \ . \tag{2.21d}$$

To determine the characteristic oscillations, we linearize Eqs. (2.21) about the uniform state $\rho = \rho_0$, $p = p_0$, $\mathbf{B} = B_0 \, \hat{e}_z$, $\mathbf{v} = \mathbf{v}_0 = 0$. We imagine this state is perturbed a very small amount, and seek an equation that describes the time evolution of these small perturbations. That is, we assume all variables can be written in the form $f = F_0 + f_1$, where $f_1 << F_0$, substitute into Eq. (2.21), and retain only terms that are linear in small quantities. The resulting system can be combined to yield a single equation for the perturbed velocity $\mathbf{v}$:

$$\rho_0 \frac{\partial^2 \mathbf{v}}{\partial t^2} = \gamma p_0 \nabla \nabla \cdot \mathbf{v} + \frac{1}{4\pi} \nabla \times \nabla \times (\mathbf{v} \times \mathbf{B}_0) \times \mathbf{B}_0 \ . \tag{2.22}$$

Since the background state is uniform, the perturbed quantities can be represented in terms of complex exponential functions. For example, we write $\mathbf{v}(\mathbf{r},t) = \tilde{\mathbf{v}} \exp[i(\mathbf{k} \cdot \mathbf{r} - \omega t)] + complex\ conjugate$, where $\tilde{\mathbf{v}}$ is a complex number. Equation (2.22) then reduces to the algebraic equation

$$\rho_0 \, \omega^2 \tilde{\mathbf{v}} = \gamma p_0 \, \mathbf{k} \mathbf{k} \cdot \tilde{\mathbf{v}} + \frac{1}{4\pi} \left[(\mathbf{k} \times \tilde{\mathbf{v}})(\mathbf{k} \cdot \mathbf{B}_0) - (\mathbf{k} \times \mathbf{B}_0)(\mathbf{k} \cdot \tilde{\mathbf{v}}) \right] \times \mathbf{B}_0 \ . \tag{2.23}$$

The three independent vectors $\mathbf{k}$, $\tilde{\mathbf{v}}$, and $\mathbf{B}_0$ appearing in Eq. (2.23) define three principal orientations for the normal modes. Three combinations are possible: all three vectors are parallel (one possibility); or, one is perpendicular to the other two (two non-trivial possibilities). Each of these cases defines a normal mode of the system; all other possible modes of oscillation can be written as a linear combination of these fundamental motions.

First, consider the case where the three vectors are parallel. These are compressional waves ($k \cdot \tilde{v} \neq 0$) propagating along the magnetic field. Then $k \times \tilde{v} = k \times B_0 = \tilde{v} \times B_0 = 0$, and Eq. (2.23) reduces to

$$\omega^2 = \frac{\gamma p_0}{\rho_0} k^2 . \tag{2.24}$$

We recognize $C_S^2 = \gamma p_0 / \rho_0$ as the square of the sound speed. This polarization thus describes the propagation of *sound waves* along the direction of B_0; these waves are unaffected by the presence of the magnetic field.

For the second case, we take k to be parallel to B_0 and perpendicular to $\tilde{v}$. These are incompressible waves ($k \cdot \tilde{v} = 0$) propagating along the magnetic field. Then $k \cdot \tilde{v} = \tilde{v} \cdot B_0 = k \times B_0 = 0$, and Eq. (2.23) becomes

$$\omega^2 = \frac{B_0^2}{4\pi\rho_0} k^2 . \tag{2.25}$$

These waves thus propagate with a speed $V_A = B_0 / \sqrt{4\pi\rho_0}$. They are called *shear Alfvén waves,* and V_A is called the *Alfvén velocity.* They are new modes introduced by the presence of the magnetic field.

The third possible polarization has B_0 perpendicular to both k and $\tilde{v}$, which are parallel to each other. These are compressional waves ($k \cdot \tilde{v} \neq 0$) propagating across the magnetic field. Then $\tilde{v} \cdot B_0 = k \cdot B_0 = k \times \tilde{v} = 0$, and Eq. (2.23) becomes

$$\omega^2 = \left(C_S^2 + V_A^2\right)k^2 , \tag{2.26}$$

so that these modes propagate at a speed faster than either the sound or Alfvén speeds. They are called *magnetoacoustic waves.* Note that when $C_S^2 \ll V_A^2$ (or $8\pi p_0 / B_0^2 \ll 1$), these modes propagate at the Alfvén speed. In this case, these modes are sometimes called *compressional Alfvén waves,* to distinguish them from the shear Alfvén waves defined in Eq. (2.25).

We note that a fourth orientation, $k \cdot \tilde{v} = k \cdot B_0 = \tilde{v} \times B_0 = 0$, results in the trivial solution $\tilde{v} = 0$.

Dimensionless Variables; the Lundquist Number

We now return to the resistive MHD model. As in viscous fluid dynamics, it is useful to introduce dimensionless variables. We choose to measure distance in units of a characteristic system size a, and the time in units of the Alfvén transit time $\tau_A = a/V_A$ where V_A is the Alfvén velocity $V_A = B_0 / \sqrt{4\pi\rho_0}$ defined in terms of characteristic value of the magnetic field, B_0, and mass density, ρ_0. Then with the transformations

$$\frac{t}{\tau_A} \to t \; , \; \frac{r}{a} \to r \; , \; a\nabla \to \nabla \; , \; \frac{v}{V_A} \to v \; ,$$

$$\frac{B}{B_0} \to B \; , \; \frac{4\pi}{c}\frac{a}{B_0} J \to J \; , \; \frac{\rho}{\rho_0} \to \rho \; ,$$

$$\frac{8\pi}{B_0^2} p \to p \; , \text{ and } \frac{\eta}{\eta_0} \to \eta \; ,$$

the model equations can be combined to yield

$$\frac{\partial \rho}{\partial t} + \nabla \cdot (\rho v) = 0 \; , \tag{2.27}$$

$$\rho\left(\frac{\partial v}{\partial t} + v \cdot \nabla v\right) = -\frac{\beta_0}{2}\nabla p + (\nabla \times B) \times B \; , \tag{2.28}$$

$$\frac{\partial p}{\partial t} = -\gamma p \nabla \cdot v - v \cdot \nabla p + \frac{2(\gamma-1)}{S\beta_0}\eta J^2 \; . \tag{2.29}$$

$$\frac{\partial B}{\partial t} = \nabla \times (v \times B) - \frac{1}{S}\nabla \times (\eta \nabla \times B) \; , \tag{2.30}$$

$$\nabla \cdot B = 0 \; . \tag{2.31}$$

Equations (2.27) through (2.31) are the usual set of resistive MHD equations. The constant $\beta_0 = 8\pi p_0 / B_0^2$ relates the normalization values for the pressure and the magnetic field. The non-dimensional parameter $S = \tau_R/\tau_A$ is the *Lundquist number*, and is a measure of the separation of the time scales characteristic of diffusive and dynamical processes. The characteristic time scale for resistive diffusion of magnetic structures of size a is given by

$$\tau_R = 4\pi a^2 / c^2 \eta_0 \; , \tag{2.32}$$

where η_0 is a characteristic value of plasma resistivity (corresponding to a characteristic temperature T_0, for example.) This is usually the longest of the characteristic time scales described by the resistive MHD equations. The fastest time scale is generally τ_A, the Alfvén time. This is the characteristic time scale for plasma dynamics. Other processes, to be discussed later, evolve on a hybrid time scale that is considerably slower than τ_A, but still much faster than τ_R. In mathematical terms, large values of S imply that Eqs. (2.27) through (2.30) are *stiff*.

We further remark that *the solutions of the resistive MHD equations in the limit* $\eta \rightarrow 0$ ($S \rightarrow \infty$) *do not converge to the solutions of the ideal* ($\eta = 0$) *MHD equations.* This is because the ideal MHD equations are a hyperbolic system and require fewer boundary conditions than the resistive MHD equations, which are of higher order in terms proportional to the resistivity. The presence of these terms relaxes certain topological constraints imposed on the solutions of the ideal MHD equations and allows for motions that are not possible when $\eta = 0$. Thus the terms proportional to η must be retained no matter how large S becomes.

The Lundquist number is also a measure of the stiffness of spatial scales. As will be described later in this chapter, the resistive MHD equations inevitably generate fine scale spatial structure as a result of their quadratic nonlinearity. This structure will continue to cascade to shorter length scales until the dissipation rate at that length scale just balances the input rate due to the cascade process. In resistive MHD, this dissipation length scale is proportional to η_0 (S^{-1}). Thus large S implies that both the time and spatial scales are widely separated.

Note that S is defined in terms of the time scales used to normalize the equations. It is not to be confused with the magnetic Reynolds' number $R_M = \tau_R v / a$, which is defined in terms of the local flow velocity v, and thermodynamic properties; S and R_M are equal only when $v = V_A$. Note also that S scales like a / η_0, so that S can become large both for fluids that are good conductors (η_0 small), and for large structures (a large.) For example, in a hot fusion plasma S is of the order of 10^6 to 10^8, while for large, cool solar coronal structures S can exceed 10^{12}.

2.2 MHD Stability

Linear Stability of Normal Modes

We have seen that the MHD equations support several normal modes, ranging from Alfvén and sound waves to resistive diffusion. These modes are constantly excited by thermal fluctuations and are always present at a low level in real systems.   Much of the motivation for studying the physics of plasmas stems from the tendency of one or more of these modes to exhibit unstable behavior in almost any non-trivial geometry.   In their most robust form these instabilities can lead to gross disruptions of magnetic configurations with a resulting release of large amounts of energy in the form of heat and radiation.   In other cases they may lead to effectively enhanced transport.   In either case they are important physical phenomena, and have been the subject of intense research for over thirty years.   The nonlinear evolution of these instabilities is responsible for plasma relaxation.

As is the case with many natural systems, a magnetized plasma tends to seek the equilibrium state of lowest potential energy consistent with existing external global constraints.   Examples of such constraints are fixed magnetic flux or total electric current.   Transitions from states of higher energy to states of lower energy are accomplished by means of instabilities, i.e., one of the ubiquitous normal modes of the system becomes unstable.   Mathematically, this comes about when non-trivial solutions of the dispersion relation Eq. (2.23) require ω to be a complex number.

A homogeneous plasma in a uniform magnetic field, discussed previously, is an example of a minimum energy state, albeit a rather uninteresting one.   In this case the magnetic flux is fixed and the current vanishes.   Any deviation from a state of uniform pressure or magnetic field increases the potential energy.   This excess potential energy is available to drive instabilities.   (Whether or not an instability actually occurs depends on the details of the system parameters; not all nonuniform states are unstable!)   The amount of this excess potential energy that is available to drive instabilities (i.e., to be converted into kinetic energy and ultimately dissipated) is called *free energy*.   In the MHD model there are two sources of free energy: nonuniform plasma pressure, and nonuniform parallel current density.   These are discussed in more detail below.

In general, to confine a hot plasma with a magnetic field implies inevitable pressure nonuniformity and non-vanishing current density.   In any experimentally interesting geometry the resulting magnetic field lines are curved in space.   In many ways such a configuration is analogous to supporting a dense fluid (the plasma) with a less dense one (the magnetic field), a situation that is known to be unstable from fluid mechanics: the system can lower its potential energy by interchanging the two fluids.   Such motions are known as *interchange modes*; their source of free energy is nonuniform plasma pressure.   (In MHD the curvature vector of the magnetic field plays a role analogous to gravity in fluid mechanics.)   However, if the magnetic field is not unidirectional, or is rigidly anchored at some point in space, then it turns out that a displacement of fluid elements that lowers the mechanical potential energy of the system in this way may also bend the magnetic field lines, thus increasing the magnetic potential energy. Instability occurs only if the *total* potential energy is lowered as a result of the displacement. Thus the onset of unstable behavior is determined at each point in the plasma by a delicate local balance between the pressure gradient and the magnetic *shear* (a measure of the local rate of change of direction of the magnetic field.) It has been found that large scale interchange modes can be completely stabilized by suitable tailoring of the shear. The pressure driven modes that remain tend to be very localized and nondisruptive; however, they may limit the amount of plasma that can be confined by a given amount of total current.

The second source of free energy available to drive instabilities arises from nonuniformities in the current density. These modes have their origin in the fact that parallel currents attract each other. Thus any perturbations to a state of nonuniform current density (as will inevitably occur due to the low level presence of normal modes) can result in unbalanced forces that tend to make the perturbations grow. Modes of this type are called *current driven modes*, and can clearly arise even in states with uniform plasma pressure.

Magnetic Shear and Singular Surfaces

As stated above, the concept of magnetic shear is important in understanding the onset and evolution of MHD instabilities. Specifically, a magnetic field is said to be sheared if its orientation in a given plane varies in the direction normal to that plane. For example, the configuration $B_y = \sin x$, $B_z = \cos x$ is sheared in the x-direction, since its orientation $\theta = \tan^{-1}(B_y/B_z)$ in the (y,z) plane is a function of x. The magnetic field lines in this system lie in the planes $x = constant$. These planes are the flux surfaces discussed in Section 1.3.

An interesting perturbation to this system may be characterized by a two-dimensional wave (propagation) vector $\mathbf{k} = k_y\,\hat{\mathbf{e}}_y + k_z\,\hat{\mathbf{e}}_z$, and can be written in the form

$$f(\mathbf{r}) = f_\mathbf{k}(x)\,e^{i\phi\,(y,z)} , \qquad (2.33)$$

where $\mathbf{r} = x\hat{\mathbf{e}}_x + y\hat{\mathbf{e}}_y + z\hat{\mathbf{e}}_z$ is the position vector, and $\phi = \mathbf{k}\cdot\mathbf{r}$ is the phase of the perturbation. [We note that if $f(\mathbf{r})$ is to represent a real quantity, as all physically realistic perturbations must, then the complex conjugate of the right-hand-side must be added to Eq. (2.33). This is implied, although not explicitly stated, throughout this presentation.] The wavefronts of this perturbation lie in the (y,z) plane, and are defined by the condition

$\phi = constant$. These wavefronts have an orientation $\psi = \tan^{-1}(-k_z/k_y)$ that is *independent* of x. They can be thought of as surfaces of constant displacement. For a given value of x, x_0 say, they will lie at an angle with respect to the orientation of the magnetic field, since in general $\psi \neq \theta$. Clearly the effect of such a displacement, which varies along a field line, is to cause the magnetic field to bend. As discussed above, such displacements tend to be stable as they result in an increase in magnetic energy. This stabilizing effect vanishes only at the special surfaces $x = x_s$ upon which $\psi = \theta$, i.e., the wavefronts of the perturbation are parallel to the magnetic field. On these surfaces the displacement does not vary along the field line. Combined with the requirement $\phi = constant$, we are led to expect that unstable displacements (should they exist) will be localized near surfaces $x = x_s$, where x_s is a root of the equation

$$F(x) = \mathbf{k} \cdot \mathbf{B}_0 = 0 \ . \tag{2.34}$$

On this surface of constant displacement, the parallel wavelength is infinite $(k_{||} = 0)$.

Surfaces on which Eq. (2.34) is satisfied are sometimes called *singular surfaces*, since the differential equation that precisely describes the normal modes is singular there. (As explained shortly, these special surfaces are also called *rational*, or *resonant*. These terms are used synonymously throughout this book.) It also turns out that the details of the stability analysis depend upon $F'(x_s)$, the local rate of change of F, with large values being stabilizing. This quantity is just what we have called the magnetic shear; hence the statement that shear can stabilize localized modes. Clearly this arises because a given displacement *away* from the $x = x_s$ plane causes more field bending in a system where F varies rapidly than it will in one where F is nearly constant.

Toroidal Pinch Configurations

It is of interest to consider the concepts discussed above for toroidal pinch configurations, such as the RFP or tokamak. For this purpose it is sufficient to consider such a plasma as a doubly periodic cylinder (r, θ, z) of radius a and length $L = 2\pi R$, where R is the major radius of the torus. (Strictly, this approximation requires that $R/a \to \infty$, or that effects due to toroidal curvature be small at finite R/a. The latter is well satisfied for the RFP; it is not for the tokamak. In either case, the approximation is completely suitable for the present heuristic discussion.) The ideas of the preceding paragraphs then apply with the *ansatz* $x \to r$, $y \to \theta$, along with the further restriction that all quantities must be single valued functions of the periodic coordinates θ and z. This restricts the components of the wave vector $\mathbf{k}$ to assume the discrete values $k_\theta = m/r$, $k_z = n/R$, where the numbers m and n

are integers and are called the *poloidal* and *toroidal mode numbers*. In such systems the unperturbed magnetic field lines are taken to lie on surfaces of constant radius. In cylindrical plasmas, these flux surfaces appear as nested, concentric cylinders. On each such surface the three dimensional field lines trace out a helix in space whose pitch is related to the ratio of the field line components B_θ/B_z, which is in general a continuous function of r. While the magnetic field components must be single valued functions of θ and z, the individual magnetic field lines need not close upon themselves after a finite number of transits of the periodic cylinder (turns around the torus), and in general fill a given flux surface ergodically, passing arbitrarily close to any point on the surface.

In light of our previous discussion, we expect instabilities may occur whenever Eq. (2.34) is satisfied. In our cylindrical coordinates this condition becomes

$$q(r) = rB_z/RB_\theta = -m/n . \qquad (2.35)$$

The function $q(r)$ is called the *safety factor*, and surfaces where Eq. (2.35) is satisfied are the resonant surfaces defined by Eq. (2.34). Modes (i.e., m, n pairs) for which Eq. (2.35) is satisfied are called *resonant modes*. (Note also that these are locations where q is a rational number; hence, the terminology *rational surface*.) Clearly, upon such surfaces the field lines close upon themselves after a finite number of toroidal transits. Also, potentially unstable displacements with mode numbers (m, n) can be avoided by constructing the magnetic geometry such that the corresponding mode rational surface does not appear in the plasma. (The effects of magnetic shear are now expressible in terms of the quantity $|(dq/dr)/q|$, with large values being stabilizing.)

Consider the consequences of these concepts for the design of toroidal plasma confinement devices. It turns out that two of the most dangerous modes for plasma confinement are the *sausage mode* (with poloidal mode number $m = 0$) and the *kink mode* (with poloidal mode number $m = 1$.) Plasma displacements associated with these modes are sketched in Figure 2-1. The sausage mode can be mitigated by requiring $q \neq 0$ in the plasma, i.e., requiring that the axial (toroidal) field not vanish. The kink mode can be avoided if $q > 1$ everywhere. (This is called the *Kruskal-Shafranov condition*.) From Eq. (2.35), and using $n = -1$, this implies that the magnetic field components must satisfy $B_\theta/B_z \leq a/R$. Devices that approximately satisfy these criteria are called *tokamaks*. There is another class of toroidal devices that satisfies neither of these criteria. These devices are called RFPs. We see later in this chapter that these plasmas are, in fact, inherently unstable to the classes of modes just described. However, their magnetic geometry is such that these instabilities generate a dynamo mechanism that sustains the

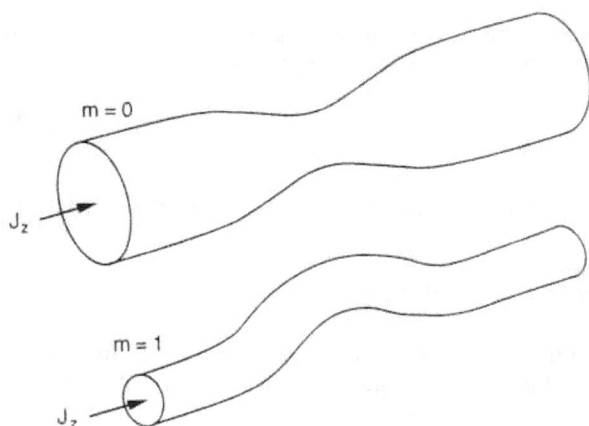

Figure 2-1. Sausage ($m = 0$) and kink ($m = 1$) modes.

discharges against resistive diffusion without terminating confinement. This is the fundamental route to the relaxed plasma state.

Resistive Instabilities

All of the previous discussion applies to perfectly conducting (zero resistivity) plasmas. The characteristic time scale associated with the normal modes of such a configuration is the *Alfvén transit time* $\tau_A = a/V_A$. Thus stable modes oscillate with frequency $\omega \approx \tau_A^{-1}$, and unstable modes grow exponentially like $e^{\gamma t}$, with characteristic growth rate $\gamma \approx \tau_A^{-1}$. For most plasmas, the Alfvén timescale is the fastest timescale supported by the MHD equations.

The addition of even the smallest amount of resistivity can completely alter the stability properties of a plasma described by the MHD model. To see that these new modes have no counterpart in ideal MHD, we introduce [*Furth*, 1969]

$$\mathbf{B} = \mathbf{B}_0(x) + \mathbf{B}_1(x,y,z,t) \tag{2.36a}$$

$$\mathbf{v} = \mathbf{v}_1(x,y,z,t) \tag{2.36b}$$

$$\mathbf{B}_0(x) = B_{z0}(x)\,\hat{\mathbf{e}}_z + B_{y0}(x)\,\hat{\mathbf{e}}_y \tag{2.36c}$$

into Eq. (2.30), assume $\nabla \cdot \mathbf{v}_1 = 0$, and ignore terms that are quadratic in quantities subscripted 1. The resulting equation can be written

$$\frac{\partial \mathbf{B}_1}{\partial t} = \mathbf{B}_0 \cdot \nabla \mathbf{v}_1 - \mathbf{v}_1 \cdot \nabla \mathbf{B}_0 + \frac{1}{S} \nabla^2 \mathbf{B}_1 \tag{2.37}$$

where constant resistivity ($\eta = 1$) has been assumed. We write first order quantities as

$$f_1(x,y,z,t) = f_1(x) \exp[\gamma t + i(k_z z + k_y y)] \quad . \tag{2.38}$$

Here we have anticipated instability and written $\omega = i\gamma$. Introducing this into Eq. (2.37), and taking the x-component of the resulting equation, we find

$$\gamma B_{x1} = iF(x)v_{x1} + \frac{1}{S}\left(\frac{d^2 B_{x1}}{dx^2} - k^2 B_{x1}\right) \tag{2.39}$$

where $\mathbf{k} = k_z \hat{\mathbf{e}}_z + k_y \hat{\mathbf{e}}_y$ is the wave vector of the perturbation and, as before, $F(x) = \mathbf{k} \cdot \mathbf{B}_0$.

   In Figure 2-2a we sketch the equilibrium field configuration, Eq. (2.36c), where we have set $B_{z0}(x) = 0$ for simplicity. The singular surface $x = x_s$, where $F(x) = 0$, thus corresponds to $B_{y0}(x) = 0$. Note there is no x-component of $\mathbf{B}$ at $x = x_s$. In Figure 2-2b we sketch a different configuration, one in which $B_{x1}(x_s) \neq 0$. Such a structure is called a *magnetic island*. We inquire into the possibility of forming the configuration of Figure 2-2b from the equilibrium of Figure 2-2a.

   From Eq. (2.39) we see that when $\eta = 0$ ($S \to \infty$), finite B_{x1} at $x = x_s$ requires v_{x1} to be infinite there. Such motions are thus prohibited by the ideal MHD equations. However, Eq. (2.39) shows that the introduction of even the smallest amount of resistivity relaxes this constraint and allows for motions that are otherwise impossible. These motions, with non-vanishing B_{x1} at $x = x_s$, are characterized by the formation of *magnetic islands* of wavelength $2\pi/k$. The presence of resistivity allows the field lines to *reconnect* across the singular surface with a *change in magnetic topology*. Modes of this type are called *resistive instabilities*. The most common of these is the *tearing mode*, so named because the magnetic field "tears" at the singular surface. A typical configuration is shown in Figure 2-3.

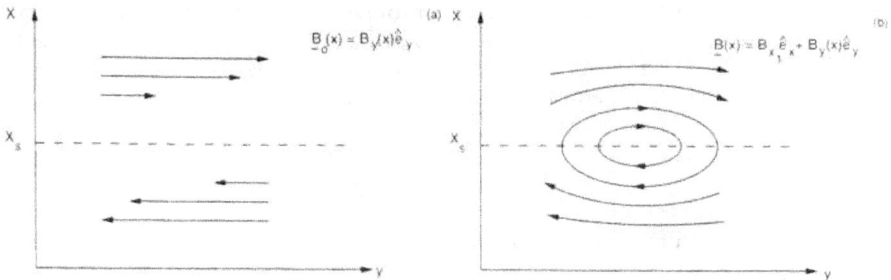

Figure 2-2 . Sketch of sheared magnetic field. a. before, and b. after reconnection.

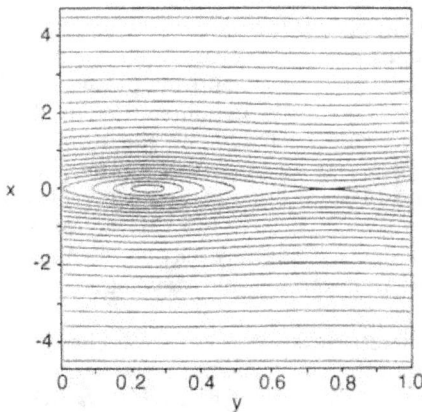

Figure 2-3. Example of a magnetic island.

The mathematical theory of resistive instabilities is well developed, and beyond the scope of this book [*Furth et al.*, 1963]. However, the length and time scales associated with resistive modes can be found by considering the basic physics underlying the reconnection process. Without loss of generality, we orient our coordinate system so that the y axis is parallel to $\mathbf{k}$, and $x = 0$ defines the singular surface. The ensuing analysis [*Furth et al.*, 1963] is then carried out in the x-y plane (y being the direction of the wave vector), where it is the vanishing of the y-component of the zero-order field that is significant (the singular surface is thus defined by $B_{y0} = 0$).

Now consider an element of fluid lying in the region $x > 0$ where $B_{y0} > 0$, and let us imagine that we can move this fluid element at velocity $\mathbf{v}_1$ without perturbing the magnetic field, i.e., for the motion we are considering $\partial \mathbf{B}/\partial t = 0$ and the perturbed electric field vanishes, or at least is electrostatic. Then from Eqs. (2.17) and (2.20) we see that there is a current density

$$\mathbf{J}_1 = \frac{\mathbf{v}_1 \times \mathbf{B}_0}{\eta c} \tag{2.40}$$

associated with this motion, with a resulting body force

$$\mathbf{F}_1 = -\frac{v_{1\perp} B_0^2}{\eta c^2} \hat{\mathbf{e}}_x \tag{2.41}$$

on the fluid element that opposes the motion. (Here, the subscript $\perp$ refers to the direction perpendicular to the zero-order magnetic field.) When $\eta \to 0$, this force can become arbitrarily large, and the plasma cannot flow separately from the field.   (This is just the usual statement that in an infinitely conducting plasma the field is "frozen in" the fluid.)   With $B_0 = B_{y0}$ (as discussed above), we find that for small but finite η this force is still large throughout most of the plasma, but can become arbitrarily weak near the singular surface, where $B_0 = 0$. If there is a driving force $\mathbf{F}_d$ that opposes the restoring force given by Eq. (2.41), it may become dominant in the region near the singular surface and an instability may develop.

Let a be a characteristic scale length for changes in the zero-order magnetic field. In light of the above discussion we expect there to be a region of flow detached from the field lines of half width εa, where $\varepsilon \ll 1$, centered about the singular surface. At the edge of this region the (as yet unspecified) driving force is approximately equal and opposite to the restoring force. Writing

$$B_0(\varepsilon a) \approx B_0'(0)\varepsilon a \tag{2.42}$$

we find that the power generated by this force is

$$P = \mathbf{F}_1 \cdot \mathbf{v} \approx \frac{v_1^2 (B_0')^2 (\varepsilon a)^2}{\eta c^2} \tag{2.43}$$

Now if we assume incompressible flow that is insignificant for $x > \varepsilon a$, we find $|v_{y1}/v_{x1}| \approx 1/\varepsilon a k \gg 1$. Then the kinetic energy of the fluid in the neighborhood of the singular layer is

$$E_K \approx \rho v_{y1}^2 \approx \frac{\rho v_{x1}^2}{k^2 (\varepsilon a)^2} \tag{2.44}$$

Equating the rate of change of this quantity to the power given by Eq. (2.43), we arrive at an expression for the thickness of the layer of detached flow (the *singular layer*)

$$\varepsilon a = \left[\frac{\gamma \rho \eta c^2}{k^2 B_0'^2} \right]^{1/4} \tag{2.45}$$

where γ is the growth rate of the instability (should one exist.) Equation (2.45) defines the extent of the region over which resistive effects are dominant.

Instability requires the presence of a driving force F_d. To find an appropriate driving force we now consider the complete form of Ohm's law within the singular layer, whose z-component is

$$\eta J_{z1} = E_{z1} + \frac{1}{c} v_{x1} B_0 \tag{2.46}$$

where the perturbed electric field is induced by the perturbed magnetic field,

$$E_{z1} \approx \frac{\gamma B_{x1}}{ck} \tag{2.47}$$

and the perturbed current is

$$J_{z1} \approx \frac{c}{4\pi k} B_{x1}'' \tag{2.48}$$

Now, by detailed consideration of the solution of the ideal MHD equations *away* from the singular layer [*Furth et al.*, 1963], it can be shown that these solutions have a singularity in the logarithmic derivative of the perturbed field *at* the singular surface that, in the limit $ka \ll 1$, is given by

$$\Delta' \equiv a \lim_{\varepsilon \to 0} \frac{B_{x1}'(\varepsilon a) - B_{x1}'(-\varepsilon a)}{B_{x1}(0)} \approx \frac{1}{ka} \tag{2.49}$$

so that we can express the second derivative of the perturbed field as

$$B_{x1}'' \approx \frac{B_{x1}'}{\varepsilon a} \approx \frac{\Delta' B_{x1}}{\varepsilon a^2} \approx \frac{B_{x1}}{\varepsilon k a^3} . \tag{2.50}$$

Then the perturbed current Eq. (2.48) becomes

$$J_{z1} = \frac{cB_{x1}}{4\pi ek^2 a^3} .$$

(2.51)

A detailed mathematical analysis shows that instability requires $\Delta' > 0$ [*Furth et al.*, 1963]. Thus stability to resistive modes can be determined by computing Δ' with the ideal MHD equations in the region away from the singular layer and examining the sign of Δ'. The determination of the actual growth rate γ requires that the full resistive MHD equations be solved.

Within the region εa of detached flow the induced electric field in Eq. (2.47) is dominant, and the ensuing Lorentz force drives the instability. Then balancing $\eta J_{z1} \approx E_{z1}$, we can determine εa from Eqs. (2.47) and (2.51). The result is

$$\varepsilon a = \frac{\eta c^2}{4\pi k \gamma a^2} .$$

(2.52)

Equating Eqs. (2.45) and (2.52) allows us to solve for the growth rate γ. We find

$$\gamma = \left[\frac{B_0^2 \eta^3 c^6}{(4\pi)^4 \rho k^2 a^{10}} \right]^{1/5}$$

(2.53)

where we have written $B_0' \approx B_0 / a$. Notice that γ scales like $\eta^{3/5}$, midway between the Alfvén (η^0) and diffusion (η^1) time scales. We also note that, from Eq. (2.45) or Eq. (2.52), εa scales as $\eta^{2/5}$.

It is instructive to write Eq. (2.53) in terms of the two basic characteristic time scales: the Alfvén time τ_A, and the resistive diffusion time τ_R. With non-dimensional wave number $\alpha = ka$, this becomes simply

$$\gamma = \alpha^{-2/5} \tau_A^{-2/5} \tau_R^{-3/5} .$$

(2.54)

This is the growth rate for tearing modes, and is valid when $\alpha < 1$. As noted above, we now see that the tearing mode grows at a rate part-way between the two fundamental time scales of the MHD equations, containing 2/5 of the Alfvén time and 3/5 of the resistive diffusion time. Normalized to the Alfvén time, the growth rate can be written in terms of the Lundquist number $S = \tau_R / \tau_A$:

$$\gamma \tau_A = \alpha^{-2/5} S^{-3/5} . \tag{2.55}$$

Recall that, for interesting plasmas, S is in the range 10^4 to 10^{12}. Thus tearing modes evolve on a time scale that can be many orders of magnitude slower than that of ideal instabilities or Alfvén waves. Yet these modes are now known to be the dominant long wavelength physical phenomena in modern fusion experiments. The ergodization of magnetic fields that can result from the formation of magnetic islands can greatly enhance thermal transport and can even induce a major disruption of the discharge. In the solar context, modes of this type may be responsible for the sudden release of energy associated with solar flares. They are also thought to play a prominent role in the dynamics of the magnetosphere. It will turn out that these modes are responsible for the RFP dynamo, and hence plasma relaxation.

Nonlinear Effects

In plasma relaxation, the system makes a transition from a given state to a preferred state. Thus, the question of the stability of a given state to small perturbations is relevant to a dynamical description of relaxation. The preceding discussion has focused on the exponential growth of these infinitesimally small perturbations. As useful as it is, the theory of the stability of solutions to the linearized MHD equations sheds no light on the amplitude to which these unstable modes can grow. For example, does a particular mode stop growing (saturate) at very low amplitude, or does it grow without bound until the system is destroyed? Or does it stop growing at some large but finite amplitude without any catastrophic consequences; and, if so, what are the properties of the final state? What happens if two or more modes are simultaneously unstable? Answering these and other similar questions requires that the nonlinear equations be solved.

The principal nonlinearities in the resistive MHD equations are quadratic products, such as $\mathbf{v} \times \mathbf{B}$. There are two ways in which these nonlinearities can affect the evolution of the system. The first is by direct *mode coupling*, and the second is by *quasilinear* modification of the background.

Mode coupling occurs because the product of two complex exponentials, such as Eq. (2.38), produces harmonics at spatial wavelengths both shorter and longer than either of the original wavelengths, corresponding to the sum and difference of the original wavenumbers: $[u_1 \exp(ik_1 x) + cc] \times [u_2 \exp(ik_2 x) + cc] = u_1 u_2 \exp[i(k_1 + k_2)x] + u_1 u_2^* \exp[i(k_1 - k_2)x] + cc$, where cc and $(\ )^*$ stand for the complex conjugate. This continual generation of new spatial frequencies (length scales) can drain energy from the unstable modes and cause them to cease growing. It is also the source of the turbulent cascade to short wavelength (large k) described in Chapter 1.

Quasilinear effects occur when a mode couples to itself to generate a difference wavenumber $k = 0$, corresponding to the background state (in the example of the previous paragraph, $k_1 = k_2, u_1 = u_2$). The resulting modification of the background ($k = 0$) in which the mode is growing can remove the original free energy source and cause saturation, even if the unstable mode continues to evolve according to the linearized equations. [The modification of the background state changes the coefficients in the linearized differential equations, such as B_0 in Eq. (2.37), and thus modifies the solution B_1 and the growth rate γ. Saturation occurs when the coefficients have been modified to the extent that $\gamma = 0$.]

Except under a few particular circumstances, the nonlinear resistive MHD equations defy analytic solution. Thus, to describe the dynamics of relaxation, the resistive MHD Eqs. (2.27-2.30) must be solved numerically as an initial value problem, generally in three space dimensions. Because of the large separation between the Alfvén, tearing, and resistive time scales, the accurate and efficient calculation of the nonlinear evolution of resistive modes presents a formidable challenge, and has motivated the development of advanced numerical methods. Our present understanding of relaxation dynamics is largely the result of the application of these methods to appropriate problems. These methods are not reviewed here; the interested reader is referred to the published literature [*Schnack et al.*, 1984, 1986, 1987]. However, the results of the application of these methods are thoroughly described in Chapters 5-8.

We remark on the nonlinear behavior of tearing modes. It turns out that, because of mathematical details, the solutions to the linearized resistive MHD equations can be divided into two classes: those in which the perturbed magnetic field is approximately constant within the singular layer [see Eq. (2.52)]; and, those in which the perturbed magnetic field varies significantly over the layer. In the jargon of resistive instabilities, these are called "constant-ψ" and "non-constant-ψ" modes, respectively. *Rutherford* [1973] showed analytically that constant-ψ tearing modes saturate nonlinearly (specifically, cease exponential growth) when the width of the associated magnetic island is comparable to the width of the singular layer εa. For parameters of interest, this is generally an exceedingly small amplitude. On the other hand, it has been shown [*Drake et al.*, 1978] that no such saturation mechanism exists for non-constant-ψ tearing modes, so that these modes can exhibit robust nonlinear behavior. These predictions have been borne out by extensive numerical calculations. In toroidal devices, non-constant-ψ tearing modes are characterized by poloidal mode number $m = 1$ (i.e., they are kink modes). We may thus expect these modes to play a special role in relaxation occurring in these configurations.

2.3 Stability Properties of the RFP

We now discuss the basic linear stability properties of the RFP, and compare them with those of the tokamak. The purpose of this discussion is twofold. First, it introduces concepts and notations that are used extensively in the following chapters, where detailed descriptions of the relaxation process in the RFP are given. Secondly, it highlights the unique features of RFP dynamics, and contrasts them with the better known properties of the tokamak. We emphasize that relaxation is generally a nonlinear process. Nonetheless, linear theory can serve as a guide to sorting out the complications introduced by the nonlinearities, and can provide a language with which to describe certain phenomena. This is discussed in detail in Chapter 4.

First, consider the differences between the RFP and tokamak q profiles. In Figure 2-4, profiles for a fully formed RFP and tokamak are shown. As mentioned previously, the tokamak q profile is a monotone increasing function of minor radius with a value on axis of about unity. The RFP profile is a monotone decreasing function of radius with a value on axis of about $a/2R$ (the exact value of $q(0)$ depends on the aspect ratio of the particular device).

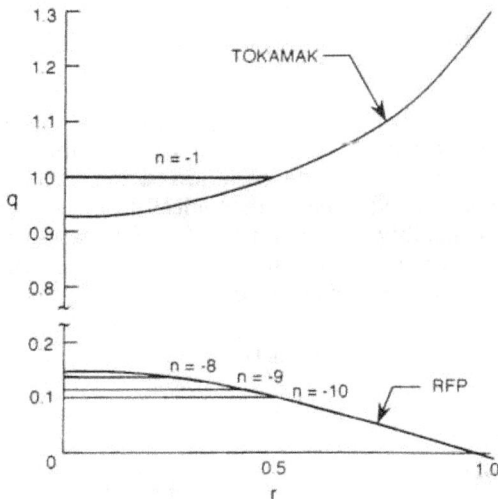

Figure 2-4. RFP and Tokamak q-profiles.

Resonant, or singular, surfaces occur when $q(r) = -m/n$, where m and n are poloidal and axial (toroidal) mode numbers [see Eq. (2.35) and the ensuing discussion.]. Here we concentrate on modes with poloidal mode number $m = 1$ because of their robust nonlinear behavior (these are non-constant-ψ modes). From Figure 2-4 we see that for the tokamak with $q \lesssim 1$ on axis there is only one such unstable mode, corresponding to $n = -1$. On the other hand, for the given RFP, when $n < -5$ there are many possible resonances in the positive q region; when $n \gg 1$ there are resonances in the negative q region as well. Therefore, in the RFP there are many possibly unstable $m = 1$ modes that can interact with each other nonlinearly; in the tokamak there is only one. The single mode in the tokamak is responsible for sawtooth oscillations. The many modes in the RFP are responsible for the dynamo. (The RFP also exhibits a sort of sawtooth oscillation. These are discussed in Chapter 7.)

In Section 2.2 we gave a heuristic discussion of the stability of the normal modes of the linearized MHD equations. There we noted that instabilities occur when small displacements away from a background state can lower the overall potential energy of the magnetoplasma system. This concept can be expressed quantitatively in terms of an *energy principle* [*Bernstein et al.*, 1958], in which the change in potential energy of the configuration is expressed in terms of the volume integral of a functional of the plasma displacement. Unstable displacements produce a negative change in potential energy. In the language of the calculus of variations, we seek to minimize the energy functional

$$W(\xi) = \frac{\pi}{2} \int_0^a \left(f \xi'^2 + g \xi^2 \right) dr \tag{2.56}$$

with respect to acceptable radial plasma displacements ξ [Eq. (2.56) is a special case of the energy principle valid in a doubly periodic cylinder; here, $v_r = \partial \xi / \partial t$, and $(\ )'$ refers to differentiation with respect to r.] These displacements are constrained to be mathematically well behaved at $r = 0$, and to vanish at $r = a$. Instability results if a function $\xi(r)$ can be found such that $W(\xi) < 0$. For $m = 1$ current driven modes, f and g are given by [*Newcomb*, 1960; *Robinson*, 1971; *Caramana et al.*, 1983]

$$f = \frac{r B_\theta^2 (nq+1)^2}{1 + k^2 r^2}, \tag{2.57a}$$

$$g = \frac{B_\theta^2 k^2 r}{\left(1 + k^2 r^2\right)^2} \left\{ (nq+1) \left[nq\left(k^2 r^2 + 3\right) + k^2 r^2 - 1 \right] \right\}, \tag{2.57b}$$

where $k = n/R$ is the axial wave number, a is the minor radius, and R is the major radius.

The first term in the integrand of Eq. (2.56) is positive definite, and is thus stabilizing. However, the second term can be negative, depending on the form of g, and hence may produce instability. From Eq. (2.57b), we see that the function g changes sign at a rational surface [where $nq(r_s) = -1$] that lies in the region of positive q: if $q > -1/n$, then $g > 0$; if $q < -1/n$, then $g < 0$. Furthermore, if, in a given radial region, $\xi = const.$ is a displacement compatible with the boundary conditions, then only the sign of $\int gdr$ over that region need be considered to determine stability [*Robinson*, 1971]. [If $\xi = const.$ is not an acceptable displacement, then the Euler differential equation associated with Eq. (2.56) must be solved to determine stability.]

Let us now apply these ideas to the tokamak. From Figure 2-4, we see that $q < 1$ and $g < 0$ inside the rational surface ($r < r_s$), while $q > 1$ and $g > 0$ outside the rational surface. Thus, an acceptable trial function that will result in $W < 0$ is $\xi = constant$ for $r < r_s$, and $\xi = 0$ for $r > r_s$. This function is shown in Figure 2-5a. Unstable $m = 1$ displacements in a tokamak are thus characterized by a radially outward shift of the plasma core, with little disturbance of the outer regions. Since $\xi = 0$ in the outer region, these modes are unaffected by the relative placement of the rational surface and the outer boundary (where we require $\xi = 0$).

Next, consider the RFP. Due to the monotone decreasing nature of the q profile, we find $g > 0$ inside the rational surface, and $g < 0$ outside the rational surface; the destabilizing influence is now in the *outer* regions of the discharge. Accordingly, an acceptable unstable trial function will have $\xi = 0$ inside the rational surface. However, we cannot simply set $\xi = constant$ for $r_s < r < a$, as ξ must vanish at the outer boundary. Thus, an acceptable displacement must peak at the rational surface, and then decrease smoothly to zero at $r = a$; such a displacement is shown in Figure 2-5b. Note that, since ξ' no longer vanishes, the stabilizing effect of the first term in the integrand of Eq. (2.56) is now felt in the outer region; this is the stabilizing effect of the conducting wall. As mentioned previously, the Euler equation must now be solved to determine stability. We also note that in the RFP it is possible for $q < -1/n$ to be satisfied everywhere in the plasma, so that the discharge can be unstable to modes that do not have a resonant surface in the plasma (such modes are said to be nonresonant from above); the destabilizing trial function then extends across the entire radius. We thus expect unstable $m = 1$ modes in the RFP to be resonant near $r = 0$, or to be nonresonant from above. These modes have a broad radial structure that extend to the wall.

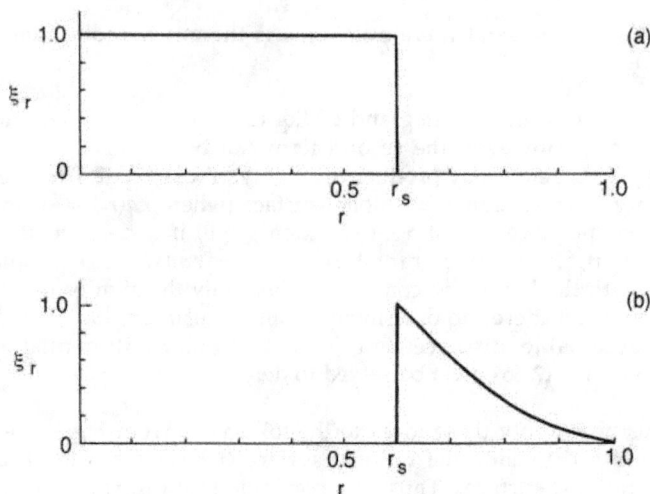

Figure 2-5a. $\xi(r)$ for tokamak; b. $\xi(r)$ for RFP.

To summarize, unstable $m = 1$ modes in the tokamak have their destabilizing terms inside the rational surface and cannot be wall stabilized, whereas for the RFP such modes have their driving terms outside the rational surface and the influence of the conducting wall is important. The tokamak can have at most one unstable $m = 1$ mode, whereas the RFP can have many, some of which need not be resonant in the plasma.

The consequences of these unstable modes for plasma relaxation has been alluded to in Chapter 1. While the discussion of this section strictly applies only when $\eta = 0$, when the resistivity is finite these modes drive the magnetic reconnection that is responsible for relaxation. The isolated mode in the tokamak is unable to provide the amount of reconnection required for Taylor's theory to be valid, while the abundance of modes in the RFP provides adequate reconnection for more complete relaxation (see Chapter 3) to occur. The dynamics of this process are discussed in Chapters 4 and 5.

2.4 The Force Free MHD Model

In Chapter 3 we will see that *equilibrium* magnetic fields for which the Lorentz force vanishes ($J \times B = 0$) play a special role in the theory of plasma relaxation; in these configurations the current density is everywhere parallel to the magnetic field. These states are said to be *force-free*. In many cases pressure plays a relatively minor role in determining the overall properties of the relaxed state. This is especially true in the RFP, where finite plasma pressure arises from small imbalances between the stresses in the poloidal and toroidal fields. Thus, the dynamics of plasmas for which $\nabla p = 0$ are of special interest in studies of relaxation, as any resulting steady states (for which $\partial/\partial t = 0$) will be force-free.

Under the foregoing assumptions, the resistive MHD equations are greatly simplified. The condition $\nabla p = 0$ now serves as the equation of state that allows closure of the system; the energy equation need not be solved. Furthermore, it is often assumed that the mass density ρ is uniform and constant. This assumption is generally made for computational convenience, although in the absence of thermodynamic evolution and pressure forces the mass density is unlikely to play an active role in determining the dynamics. (Except for compressible effects resulting from finite $\nabla \cdot \mathbf{v}$, the density behaves as a passively advected scalar.) It also proves computationally convenient to introduce a vector potential $\mathbf{A}$, such that $\mathbf{B} = \nabla \times \mathbf{A}$, and to use a gauge condition in which the scalar potential vanishes. Then the resistive MHD Eqs. (2.27-2.30) become

$$\frac{\partial \mathbf{A}}{\partial t} = -\mathbf{E}, \tag{2.58a}$$

$$\mathbf{E} = -\mathbf{v} \times \mathbf{B} + \frac{\eta}{S} \mathbf{J}, \tag{2.58b}$$

$$\rho_0 \left(\frac{\partial \mathbf{v}}{\partial t} + \mathbf{v} \cdot \nabla \mathbf{v} \right) = \mathbf{J} \times \mathbf{B}. \tag{2.58c}$$

This form is especially useful because Eq. (2.58a) allows convenient formulation of boundary conditions in which applied voltages are specified.

The force-free model Eq. (2.58) has been used extensively in computational studies of plasma relaxation. Many of these results are described in Chapters 5-8.

2.5 The Role of Numerical Simulation

A goal of this book is to demonstrate how long wavelength, low frequency, coherent MHD fluctuations can produce plasma relaxation. We shall see that these motions are the nonlinear state of linearly unstable normal modes of the system. Through their nonlinear dynamical interaction they produce the magnetic field configurations and fluctuations that are observed experimentally. The inherent nonlinearity of the process renders analytic treatment virtually impossible. As a result, numerical simulation has played a central role in determining the picture that is presented.

In principle, for the RFP the problem to be solved is simple. An initial value problem is posed in a periodic cylinder (as is appropriate to lowest order in the toroidal inverse aspect ratio) in which the initial conditions consist of some representative large amplitude, large scale magnetic field that depends only on radius, plus a small amplitude, three dimensional, spatially random (but divergence-free) magnetic field that represents a perturbation to the large amplitude background. This perturbation may be thought of as resulting from thermal fluctuations, for example, and excites all of the normal modes of the system. The boundary conditions may be those of a perfect electrical conductor, or of a resistive shell. The nonlinear, three-dimensional, resistive MHD equations are then integrated forward in time. Any normal modes that are unstable grow to finite amplitude, interact with each other, and saturate nonlinearly. Since we anticipate that, at least in the RFP, these modes are current driven, the force-free MHD model presented in Section 2.4 is often used. The time-asymptotic solution should exhibit the properties of plasma relaxation.

Now consider the difficulties inherent in such an approach. We have shown that the resistive MHD equations admit several characteristic timescales. The fastest of these is the Alfvén time τ_A; the slowest is the resistive diffusion time τ_R. The ratio of these is the Lundquist number $S = \tau_R / \tau_A$. For real plasmas, S is in the range of 10^4 to 10^{12}, so that these timescales are widely separated. We see that relaxation occurs on a third timescale set by resistive instabilities. These normal modes evolve on a timescale that is intermediate between fast Alfvén waves and resistive diffusion; thus $\tau_{RELAX} = \tau_A^{1-\nu} \tau_R^{\nu}$, where $0 < \nu < 1$. Clearly, $\tau_{RELAX} \gg \tau_A$.

Any numerical simulation proceeds at a finite time step Δt, and with finite spatial resolution $\Delta x \ll L$, where L is a characteristic length scale for the configuration. For studies of plasma relaxation, the system should be integrated for many relaxation times. We would thus like to have $\Delta t \lesssim \tau_{RELAX} = L/v$. (Here v is a characteristic velocity that is much less than the Alfvén velocity $v_A = L/\tau_A$.) However, standard numerical methods require that Δt be limited for reasons of numerical stability to be less than $\Delta x / v_A$. Thus, we require $\Delta t \ll \tau_{RELAX}$, making the use of standard algorithms for realistic plasmas prohibitively expensive.

Progress in numerical studies of plasma relaxation has therefore required the development of advanced algorithms for time integration. These semi-implicit methods [Harned and Kerner, 1985, 1986; Harned and Schnack, 1986; Schnack et al., 1987] allow Δt to be determined for reasons of accuracy alone, so that calculations can proceed on the relaxation timescale. Even then, a typical calculation may take tens of hours of CPU time on the fastest available supercomputer. Most of the results presented in this chapter could not have been obtained without the development of such techniques. The details of these numerical methods are beyond the scope our presentation; rather, the reader is referred to the referenced literature.

We further remark on the spatial resolution required in these calculations. It is well known [*Landau and Lifshitz*, 1959] that the number of degrees of freedom per spatial dimension for a nonlinear turbulent system is approximately equal to the Reynolds' number; in our case this number is the Lundquist number S. In principle, it is this number of modes that should be resolved numerically (i.e., $\Delta x / L < S^{-1}$). Clearly, *fully turbulent numerical MHD simulations of naturally occurring and laboratory plasmas are beyond the capability of present computing technology.* However, long wavelength motions *can* be accurately resolved. Most of the calculations to be reported in this book use a combination of finite differences in the radial dimension and finite-Fourier transforms in the two periodic dimensions. Typical resolutions (after numerical dealiasing) are 100 radial mesh points, five poloidal Fourier modes, and 170 axial (toroidal) Fourier modes. The inevitable buildup of energy at the small length scales is controlled by some form of artificial viscosity.

The results of this type of calculation are presented in Chapters 5 through 8. From these it becomes clear that relaxation in certain configurations can result from long wavelength modes. However, since small scale turbulence at relatively high Reynolds' number cannot be simulated, its effect cannot be directly addressed by the methods discussed here. The clarification of the role of turbulence in plasma relaxation depends on advances in large scale computing technology. Perhaps the next generation of supercomputers based on massively parallel architectures will be useful for this task.

CHAPTER 3
TAYLOR'S THEORY OF PLASMA RELAXATION

In Chapter 2 the elements of the resistive MHD model were presented. In this chapter we begin the application of this model to a description of plasma relaxation.

It is observed that many magnetoplasma systems tend to naturally exist in states that are relatively independent of the initial conditions, or of the way in which the system is initiated. The properties of the system seem to be completely determined by the boundary conditions and a few global parameters, such as flux, current, or applied voltage. For example, if a series of experiments is performed with the same plasma density, voltage, and flux, approximately the same field profiles and plasma parameters are achieved, even though the precise details of the way in which the plasma initially fills the discharge tube may differ significantly between experiments. These states are achieved in a finite length of time after the system is initiated; if the system is then disturbed, it returns to the preferred state, again in a finite time. This process is referred to as *plasma relaxation*, and the states as *relaxed states*. The relaxed states appear to be reproducible down to the details of the spatial distribution of current and magnetic field. Dynamical properties, such as characteristic fluctuations, are also relatively invariant across realizations. Such configurations cannot result from force balance or stability considerations alone, as many different examples of stable equilibria consistent with a given set of boundary conditions are known to exist. Clearly, some other processes are at work.

A theory of relaxed states was first given by J. B. Taylor in 1974. This theory accounts, at least qualitatively (and often quantitatively) for many of the observed properties of laboratory and astrophysical plasmas. Mathematically, the theory is based on a variational principle in which the potential energy of the magnetoplasma system is minimized subject to the constraint that a certain volume integral, called the *magnetic helicity*, is invariant. This constraint is related to a global topological property of the magnetic field. The theory is thus thermodynamic in the sense that the details of the underlying dynamical processes responsible for plasma relaxation do not appear explicitly; rather, the macroscopic properties of the equilibrium states that result from the microscopic dynamics are computed. Taylor's theory has been particularly successful in providing a framework for describing the operation of the RFP [*Bodin and Newton*, 1980]. In this chapter, we present this theory, along with a discussion of its possible shortcomings. We also present examples of the successful application of the theory to RFP and multipinch plasmas.

3.1 The Constraints of Ideal MHD

As discussed in Chapter 2, it is a well established physical principle that an isolated system tends to seek a state of minimum potential energy. An example of such a system is a plasma permeated by a magnetic field and surrounded by a boundary that is a perfect electrical conductor. Then the tangential electric field, as well as the normal magnetic field, must vanish at the system boundary. Consequently, there can be no inward or outward Poynting flux, $S = E \times B$, and no electromagnetic energy can enter or leave the system. We also assume that the system is thermally isolated from the outside environment. The potential energy for such a system is given by

$$W = \int_{V_0} \left(\frac{B^2}{8\pi} + \frac{p}{\gamma - 1} \right) dV . \tag{3.1}$$

Here, B is the magnetic field strength, p is the plasma pressure, and γ is the ratio of the specific heat at constant pressure to the specific heat at constant volume. A direct minimization of Eq. (3.1) with respect to arbitrary variations in the magnetic field and pressure yields the uninteresting state $W = 0$; clearly some physically relevant constraints must be applied to the minimization process. In the presence of a perfectly conducting outer boundary, one such constraint is that the toroidal flux

$$\Phi_z = 2\pi \int_0^a B_z r dr , \tag{3.2}$$

be fixed. Such a minimization yields a state of zero current; further constraints are needed. It will now be shown that, for a perfectly conducting plasma, there are, in fact, an infinite number of such constraints.

The Woltjer Constraints

Following *Moffatt* [1978], we consider a closed (unknotted) curve C spanned by a surface S, with normal unit vector n. Then the flux Φ of B through S is defined as

$$\Phi = \int_S B \cdot n dS = \oint_C A \cdot dl , \tag{3.3}$$

where $\mathbf{A}$ is the vector potential, and the line integral is taken in the usual right-handed sense. A *flux tube* is defined as the volume swept out by all the field lines passing through a given closed curve C. (Usually, C is taken to be infinitesimally small.) Since $\nabla \cdot \mathbf{B} = 0$, these tubes will either close upon themselves, or will fill all space ergodically. On the lateral sides of each tube [*not* the cross-sectional surface S defined in Eq. (3.3)], $\mathbf{B} \cdot \mathbf{n} = 0$, by construction. A finite volume will contain an infinite number of such flux tubes. Clearly, Φ is constant along each flux tube.

Now consider the infinite set of integrals, first defined by *Woltjer* [1958],

$$K_l = \int_{V_l} \mathbf{A} \cdot \mathbf{B} dV \ , \ l = 1, 2, \ldots, \infty \ , \tag{3.4}$$

where V_l is the volume of the lth flux tube, whose closed, bounding surface is S_l. The time derivative of Eq. (3.4) can be written as

$$\frac{dK_l}{dt} = -\int_{V_l} \mathbf{E} \cdot \mathbf{B} dV - \int_{V_l} \mathbf{A} \cdot \nabla \times \mathbf{E} dV + \oint_{S_l} (\mathbf{A} \cdot \mathbf{B}) \mathbf{v} \cdot \mathbf{n} dS \ , \tag{3.5a}$$

$$= -2 \int_{V_l} \mathbf{E} \cdot \mathbf{B} dV + \oint_{S_l} \left[(\mathbf{A} \times \mathbf{E}) \cdot \mathbf{n} + (\mathbf{A} \cdot \mathbf{B})(\mathbf{v} \cdot \mathbf{n}) \right] dS \ , \tag{3.5b}$$

where we have used $\partial \mathbf{A} / \partial t = -\mathbf{E}$, $\mathbf{B} = \nabla \times \mathbf{A}$, and Gauss' theorem. The last term on the right-hand-side of Eq. (3.5a) arises because we have allowed the surface of the flux tube to move with velocity $\mathbf{v}$. In a perfectly conducting fluid, $\mathbf{E} = -\mathbf{v} \times \mathbf{B} + \nabla \phi$, where $\mathbf{v}$ is the local velocity and ϕ is the scalar potential. Then, using $\mathbf{n} \cdot \mathbf{B} = 0$ on S_l, Eq. (3.5b) becomes

$$\frac{dK_l}{dt} = -2 \int_{V_l} \mathbf{B} \cdot \nabla \phi dV - \oint_{S_l} (\nabla \phi \times \mathbf{A}) \cdot \mathbf{n} dS \ . \tag{3.6}$$

The first term on the right-hand-side of Eq. (3.6) can be written as

$$\int_{V_l} \mathbf{B} \cdot \nabla \phi dV = \int_{V_l} \left[\nabla \cdot (\phi \mathbf{B}) - \phi \nabla \cdot \mathbf{B} \right] dV \tag{3.7a}$$

$$= \oint_{S_l} \phi \mathbf{B} \cdot \mathbf{n} dS = 0 \ , \tag{3.7b}$$

while the second term becomes

$$\oint_{S_l} (\nabla\phi \times \mathbf{A}) \cdot \mathbf{n}dS = \int_{V_l} \nabla \cdot (\nabla\phi \times \mathbf{A}) \, dV \qquad (3.8a)$$

$$= \int_{V_l} \mathbf{B} \cdot \nabla\phi dV = 0. \qquad (3.8b)$$

Thus each of the integrals K_l is invariant in a perfectly conducting plasma. This result, first obtained by *Woltjer* [1958], depends only on the boundary condition $\mathbf{n} \cdot \mathbf{B} = 0$, and on the relationship $\mathbf{E} = -\mathbf{v} \times \mathbf{B} + \nabla\phi$.

The Topological Properties of the Woltjer Constraints

We are thus motivated to seek states that minimize W subject to the infinity of constraints $K_l = constant$, for $0 \leq l \leq \infty$. Before proceeding, however, we remark on the physical significance of these constraints, as originally given by *Moffatt* [1969, 1978], and later by *Berger and Field* [1984], and by *Taylor* [1986]. Consider two infinitesimal flux tubes that follow two closed space curves C_1 and C_2, with magnetic fluxes Φ_1 and Φ_2, and volumes V_1 and V_2, as shown in Figure 3-1. These flux tubes link each other once.

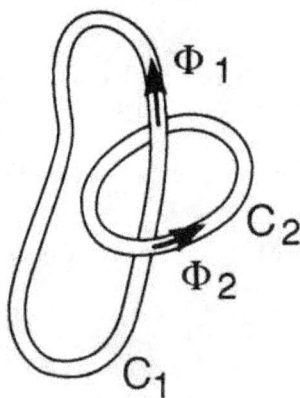

Figure 3-1. Sketch of linked flux tubes.

Then, for the first flux tube, V_1, we can write $B dV = \mathbf{B} \cdot \hat{n} dS\, dl = \Phi_1 dl$, and hence

$$K_1 = \int_{V_1} \mathbf{A} \cdot \mathbf{B} dV = \Phi_1 \oint_{C_1} \mathbf{A} \cdot dl = \Phi_1 \Phi_2 , \qquad (3.9)$$

while, for the second flux tube, V_2, we have

$$K_2 = \int_{V_2} \mathbf{A} \cdot \mathbf{B} dV = \Phi_2 \oint_{C_2} \mathbf{A} \cdot dl = \Phi_1 \Phi_2 . \qquad (3.10)$$

Thus, K_1 and K_2 measure the linkage of the two tubes of flux. If the tubes were not interlinked, the line integrals would vanish, as would K_1 and K_2; if they had been linked N times, we would find $K_1 = K_2 = \pm N \Phi_1 \Phi_2$, with the sign determined by the "handedness" of the linkage. A single knotted flux tube with flux Φ, and a flux tube with two twists, have also the same helicity. The integrals K_l are thus seen to measure a topological property of the field. The invariance of the K_l with respect to ideal MHD motions (for which $\mathbf{E} = - \mathbf{v} \times \mathbf{B} + \nabla \phi$) is another expression of the well known property of integrity of flux tubes, as derived in Chapter 2: they cannot be broken or reconnected. The topological complexity of the field, once initially established, is retained for all time.

3.2 Energy Minimization with the Constraints of Ideal MHD

We now continue with the minimization of W. We have just shown that ideal MHD, in which $\mathbf{E} = -\mathbf{v} \times \mathbf{B} + \nabla \phi$, implies that the K_l are constant. We may thus seek to directly minimize W with respect to this infinity of constraints. This requires an extension of the usual Lagrange multiplier technique [*Taylor*, 1986]. However, it can be shown [*Freidberg*, 1987] that this minimization procedure is identically equivalent to the unconstrained minimization of W under the assumption of ideal MHD. (Thus, the statement that W is minimized while the K_l remained fixed is equivalent to the assumption of ideal MHD, and *vice versa*: each implies the other.) Since the latter is mathematically simpler, we adopt this approach. Furthermore, we assume that the internal energy density is much less than the magnetic energy density, or $\beta = p/B^2 \ll 1$. Then the minimization of W reduces to the minimization of the magnetic energy. We discuss this point further later in this chapter.

To perform the minimization, we imagine an infinitesimal displacement of the plasma away from some static background state, and then calculate the resulting change in energy. Setting this energy change to zero determines a differential equation that must be satisfied by the background magnetic field. The imagined displacement is arbitrary; its effect on the magnetic field must be consistent with the constraints of ideal MHD, which can be summarized as $\nabla \cdot \mathbf{B} = 0$, $\mathbf{E} = -\mathbf{v} \times \mathbf{B} + \nabla\phi$. These allow us to express the change in the magnetic field in terms of the imagined displacement.

We now consider the effect on $\mathbf{B}$ of an arbitrary infinitesimal plasma displacement $\boldsymbol{\xi}$. By definition, this displacement gives rise to an infinitesimal change in velocity $\delta\mathbf{v} = \partial\boldsymbol{\xi}/\partial t$. Then, with $\eta = 0$, the resulting infinitesimal change in the magnetic field is

$$\delta\mathbf{B} = \nabla \times (\boldsymbol{\xi} \times \mathbf{B}) .   \tag{3.11}$$

Note that this guarantees that $\nabla \cdot \delta\mathbf{B} = 0$. Then the variation of W (with $p = 0$) is

$$\delta W = \int \mathbf{B} \cdot \nabla \times (\boldsymbol{\xi} \times \mathbf{B})\, dV = \int \boldsymbol{\xi} \cdot (\mathbf{B} \times \nabla \times \mathbf{B})\, dV .   \tag{3.12}$$

Since the displacement $\boldsymbol{\xi}$ is arbitrary, the energy is minimized ($\delta W = 0$) only if $\mathbf{B} \times \nabla \times \mathbf{B} = 0$, or

$$\nabla \times \mathbf{B} = \mu(\mathbf{r})\mathbf{B} ,   \tag{3.13a}$$

$$\mathbf{B} \cdot \nabla\mu = 0 .   \tag{3.13b}$$

Equation (3.13b) follows from Eq. (3.13a), along with $\nabla \cdot \mathbf{B} = 0$. Fields that satisfy these relations are called *force-free*, since they are equivalent to the vanishing of the Lorentz force. This result is not surprising for a low β plasma since $\mathbf{J} \times \mathbf{B} = 0$ is the condition for equilibrium force balance in the absence of pressure forces. Note that, from Eq. (2.18) and Eq. (3.13a), μ can be expressed in terms of the parallel current density:

$$\mu(\mathbf{r}) = \frac{\mathbf{J} \cdot \mathbf{B}}{B^2} .   \tag{3.14}$$

It is important to note that we are not free to arbitrarily choose the scalar function $\mu(\mathbf{r})$. According to Eq. (3.13b), μ is constant along field lines (or infinitesimal flux tubes). Recall that ideal MHD implies that we can associate an integral invariant K_l with each such field line (or flux tube), a different K_l for each field line. We can thus associate a different value of μ with each K_l, its value being determined by the initial configuration of the system. The function $\mu(\mathbf{r})$ is thus directly related to the details of the initial conditions; any system satisfying Eq. (3.13) has memory of how it was prepared. But this contradicts the properties of relaxed states, which are observed to be independent of initial conditions.   Thus, in some manner, ideal MHD over-constrains the system; a smaller set of invariants must be sought.

3.3 The Effect of Plasma Resistivity

The problem of over-constraint arises from the assumption of ideal MHD, or infinite electrical conductivity. As the name implies, ideal MHD is an idealization of the observation that real plasmas are, in fact, very good conductors. However, they are not perfect: all real plasmas have some small electrical resistivity. This resistivity is not large enough to significantly affect many plasma motions, such as Alfvén waves or fast growing instabilities; ideal MHD is a good approximation for the study of these phenomena. However, in ideal MHD the magnetic field is co-moving with the fluid; since fluid elements retain their integrity, so must magnetic field lines and flux tubes. It is just this topological property that is expressed by the infinite set of invariants K_l.

The effects of finite resistivity are introduced into the theory through Ohm's law, which becomes in nondimensional form

$$E = -v \times B + \frac{\eta}{S} J + \nabla \phi, \qquad (3.15)$$

where S is the Lundquist number.   The new term proportional to the normalized plasma resistivity η allows for non-potential contributions to the electric field parallel to B. The first term on the right-hand-side of Eqs. (3.5a,b) no longer vanishes, with the consequence that the K_l are no longer conserved; rather, they decay at a rate proportional to the resistivity. The principal effect of even the smallest amount of resistivity is to relax the topological constraints of ideal MHD, and allow for the breaking and reconnection of flux tubes. Magnetic field lines no longer maintain their integrity from one instant to the next. We would thus appear to be back where we started, with no constraints on W except the total flux.

Taylor's Conjecture

A way out of this dilemma was proposed by *Taylor* [1974, 1986]. He envisioned a slightly resistive plasma whose boundary was a rigid, perfect conductor. He then argued that, while *each* of the topological invariants K_l ceased to be relevant, the *sum* of all the invariants (the integral of $A \cdot B$ over the entire plasma volume V_0) would be independent of topological considerations and the need to identify field lines, and would thus remain a good invariant. Thus, one should minimize the energy subject to the constraint that the *total magnetic helicity*

$$K_0 = \int_{V_0} \mathbf{A} \cdot \mathbf{B} dV \tag{3.16}$$

remain constant.

For such a theory to be useful, it must be gauge invariant, i.e., the results must not change under the transformation $\mathbf{A} \to \mathbf{A} + \nabla \chi$. Introducing this into Eq. (3.16), we find

$$K_0 = \int_{V_0} \mathbf{A} \cdot \mathbf{B} dV + \oint_{S_0} \chi \mathbf{B} \cdot \mathbf{n} dS , \tag{3.17}$$

where S_0 is the closed surface bounding V_0. Thus, gauge invariance requires that the boundary condition $\mathbf{B} \cdot \mathbf{n} = 0$ be specified on S_0.

3.4 Energy Minimization with the Global Helicity Constraint

We now proceed with the minimization. Using the method of Lagrange multipliers, we minimize the integral

$$I = W - \lambda K_0 , \tag{3.18}$$

where λ is a *constant*. We find

$$\delta W = \int_{V_0} (\nabla \times \delta \mathbf{A}) \cdot \mathbf{B} dV \tag{3.19a}$$

$$= \int_{V_0} \delta \mathbf{A} \cdot \nabla \times \mathbf{B} dV + \oint_{S_0} \delta \mathbf{A} \times \mathbf{B} \cdot \mathbf{n} dS , \tag{3.19b}$$

and

$$\delta K_0 = \int_{V_0} (\delta \mathbf{A} \cdot \mathbf{B} + \mathbf{A} \cdot \nabla \times \delta \mathbf{A}) dV \tag{3.20a}$$

$$= 2\int_{V_0} \delta \mathbf{A} \cdot \mathbf{B} dV + \oint_{S_0} \delta \mathbf{A} \times \mathbf{A} \cdot \mathbf{n} dS , \tag{3.20b}$$

so that the variation of I becomes

$$\delta I = \int_{V_0} \delta \mathbf{A} \cdot (\nabla \times \mathbf{B} - 2\lambda \mathbf{B})\, dV + \oint_{S_0} \delta \mathbf{A} \times (\mathbf{B} - \lambda \mathbf{A}) \cdot \mathbf{n} dS . \qquad (3.21)$$

The gauge invariance of the theory follows immediately from Eq. (3.19a) and Eq. (3.20a), where it is seen explicitly that $\delta I = 0$ if $\delta \mathbf{A} = \nabla \chi$, and $\mathbf{n} \cdot \mathbf{B} = 0$ on S_0. The surface term in Eq. (3.21) can be made to vanish if S_0 is a perfect conductor [Taylor, 1986], for then the vanishing of the tangential electric field implies that the tangential component of $\delta \mathbf{A}$ must be the gradient of a scalar. We can then introduce the gauge transformation $\delta \mathbf{A} = \delta \mathbf{A}^* + \nabla \chi$, and choose $\mathbf{n} \times \delta \mathbf{A}^* = 0$ on S_0; the contributions from $\nabla \chi$ vanish as before, and the surface contributions from $\delta \mathbf{A}^*$ now vanish identically. The minimization thus yields the condition

$$\nabla \times \mathbf{B} = \mu \mathbf{B} , \qquad (3.22)$$

where $\mu = 2\lambda$ is now a *constant*. Relaxed states that satisfy Eq. (3.22) differ significantly from those defined by Eqs. (3.13a,b) in that they depend only on the single constant μ, independent of the initial conditions. We shall show that μ can be related to externally controllable parameters.

Validity of Taylor's Conjecture

We have just seen that minimizing W subject to the constraint that K_0 is invariant leads to relaxed equilibrium states. We shall see in Chapters 4 and 5 that the states defined by Eq. (3.22) are in relatively good agreement with observations. However, from Eqs. (3.5a,b), we see directly that, under the assumptions of the theory, K_0 is *not* invariant; it decays at a rate

$$\frac{dK_0}{dt} = -2 \int_{V_0} \frac{\eta}{S} \mathbf{J} \cdot \mathbf{B} dV \qquad (3.23)$$

The decay rate for the energy is formally of the same order in η:

$$\frac{dW}{dt} = -\int_{V_0} \frac{\eta}{S} J^2 dV \qquad (3.24)$$

One may then wonder why the theory works at all.

The key to understanding the applicability of Taylor's conjecture lies in the time scales associated with the dynamics controlling the relaxation process. The relaxed states defined by Eq. (3.22) occur only in a time asymptotic sense: they are the end result of some characteristic plasma dynamics. To be of practical use, the theory does not require that the magnetic helicity K_0 be *absolutely* invariant; it only requires that the dynamical processes responsible for the relaxation dissipate energy faster than they dissipate helicity. Thus, it is the *relative* invariance of helicity with respect to energy that is important. The present theory does not specify these dynamical processes, other than they be describable by resistive MHD.

Taylor [1974, 1986] envisioned the relaxation process as occurring in a relatively steady fashion as a result of small scale turbulence, with $S \gg 1$. For such a state, we can imagine the magnetic field being Fourier decomposed as

$$\mathbf{B} \approx \sum_k \mathbf{b_k} \exp(i\mathbf{k} \cdot \mathbf{r}) . \tag{3.25}$$

Substituting this expression into Eq. (3.23) and Eq. (3.24), we find that the decay rates for helicity and energy scale as

$$\frac{dK_0}{dt} \approx -\frac{2\eta}{S} \sum_k k b_k^2 , \tag{3.26}$$

$$\frac{dW}{dt} \approx -\frac{\eta}{S} \sum_k k^2 b_k^2 . \tag{3.27}$$

From Eq. (3.27), energy dissipation is $O(1)$ at scale lengths for which $k \sim S^{1/2}$. But from Eq. (3.26), at these scale lengths, helicity dissipation is only $O(S^{-1/2}) \ll 1$. Thus, from this heuristic argument, we may expect that small scale turbulence may dissipate energy at a greater rate than helicity.

Unfortunately, in the RFP there is no experimental evidence that relaxation is produced by small scale turbulence. The dominant magnetic fluctuations associated with the relaxation process appear to have global, long wavelength structure. This view is supported by extensive numerical simulations, which show that relaxation is produced by the nonlinear interaction of long wavelength instabilities. (Many of these results will be described in detail in Chapter 5.) Nonetheless, these simulations display the same relative invariance of helicity with respect to energy as conjectured by Taylor.

Some insight into how helicity is conserved by these long wavelength modes can be gained [*Battacharjee et al.*, 1980; *Battacharjee and Dewar*, 1982; *Hameiri and Hammer*, 1982; *Caramana et al.*, 1983] by considering the approximate invariants associated with the global magnetic reconnection resulting from long wavelength resistive tearing modes [*Kadomtsev*, 1975], having poloidal mode number m and toroidal mode number n. These instabilities evolve on time scales that are $O(S^{-\nu})$, $0 < \nu < 1$ [*Furth et al.*, 1963] (see Chapter 2), and are thus much slower than ideal MHD oscillations, but much faster than resistive diffusion. The geometry is assumed to be topologically a torus, or a periodic cylinder, as is appropriate for the RFP. Associated with each such mode is a helical flux function, which can be written in terms of the poloidal flux ψ and the toroidal flux ϕ as $\chi_{m,n} = (m/n)\psi - \phi$. We then note that, as a generalization of Eq. (3.4), the integral $\hat{K} = \int F(\mathbf{r},t)\mathbf{A} \cdot \mathbf{B}dV$ is also an invariant in ideal MHD providing the function $F(\mathbf{r},t)$ is co-moving with the fluid and constant along field lines, i.e., $\partial F / \partial t = -\mathbf{v} \cdot \nabla F$, and $\mathbf{B} \cdot \nabla F = 0$. [The invariance of $\hat{K}$ follows in a straightforward manner by following steps analogous to Eq. (3.5) through Eq. (3.8).] Any functional $F(\psi,\phi)$ of the fluxes satisfies these criteria; in particular, the infinite set of integrals

$$K_\alpha(m,n) = \int_{V_0} \chi_{m,n}^\alpha \, \mathbf{A} \cdot \mathbf{B}dV \, , \quad \alpha = 0,1,2,\dots , \qquad (3.28)$$

is approximately invariant during the reconnection process. Each such mode is associated with a particular symmetry, or pitch, proportional to the ratio m/n; there is a different set of invariants for each possible mode. However, the only invariant common among *all* possible modes is K_0, the helicity invariant of Taylor. This is also the only invariant independent of the ratio m/n, and hence may be the only one preserved when all modes are present, as is likely to occur in a realistic three-dimensional situation.

Properly Defined Helicity

In toroidal or periodic cylindrical geometry the exact definition of the integral, K_0 , used in the constraint condition must be modified. This is because in these geometries, when a toroidal (or axial, in periodic cylindrical geometry) voltage is applied, there is an additional contribution to the magnetic helicity that arises because of the linkage between magnetic flux through the center (i.e., the "hole") of the torus, due to the applied voltage, and the toroidal flux interior to the plasma. This contribution is a result of externally imposed conditions and the configuration geometry, and is not related to the structure of the magnetic field within the plasma. It is removed by redefining the constraint integral to be [*Bevir and Gray*, 1980; *Taylor*, 1980]

$$K_u = \int_{V_0} \mathbf{A} \cdot \mathbf{B}dV - \oint \mathbf{A} \cdot d\mathbf{l}_1 \oint \mathbf{A} \cdot d\mathbf{l}_2 \, , \qquad (3.29)$$

where the last term is the product of loop integrals the long and short way around the toroidal boundary. When the boundary is a complete perfect conductor (without a gap to allow the external imposition of a toroidal voltage) these integrals are constants, and the theory proceeds as before. *Taylor* [1986] has pointed out that Eq. (3.29) is required for gauge invariance when the potential χ is multivalued, as may occur in toroidal geometry. For the remainder of our discussion, we shall assume that the proper definition of helicity has been used. Consequently, we drop the subscript u, and refer only to K.

The Role of Plasma Pressure in Taylor's Theory

Before describing some detailed predictions of the theory, it is useful to remark on the role played by the plasma pressure, which has been ignored up to now. Under the assumption of ideal MHD, one can carry out the minimization of W including the pressure. The variation in pressure can be related to the imagined displacement through the ideal MHD energy equation. Then, instead of the force-free result obtained in Eqs. (3.13a,b), one finds the usual equilibrium result $\nabla p = \mathbf{J} \times \mathbf{B}$. The pressure becomes constant along flux tubes, but can vary from tube to tube in a way determined by the initial conditions. There are again many possible equilibrium states associated with a given set of externally applied parameters. However, when finite resistivity is introduced, the breaking and reconnection of field lines associated with the relaxation process allows the pressure to equilibrate over the entire plasma volume. Thus relaxed states with uniform pressure are obtained even when β is not small. The magnetic field is governed by Eq. (3.22). Formally, the variation in the pressure is now independent of the variation in the vector potential; upon performing a separate minimization, this leads to $\nabla p = 0$.

While the preceding remarks may be persuasive, a direct proof of Taylor's conjecture, or the exact circumstances under which it is applicable, is still lacking. It is therefore not surprising that several of its consequences are not borne out exactly by either experiment or numerical simulation. Nonetheless, the theory has been extremely useful for understanding the global properties of many plasma experiments, and its underlying assumptions have been confirmed by detailed numerical simulation.

3.5 Predictions of the Theory

The Reversed-Field Pinch

We now consider a specific application to the RFP. For simplicity, we consider the toroidal geometry to be approximated by a periodic cylinder ($0 \leq r \leq a$, $0 \leq \theta \leq 2\pi$, $0 \leq z \leq 2\pi R$, where a is the minor radius of the torus and R is the major radius of the torus). We consider solutions of Eq. (3.22) that have $B_r(a) = 0$, consistent with the required boundary condition $\mathbf{n} \cdot \mathbf{B} = 0$ on

S_0. Following *Freidberg* [1987], we take the z-component of the curl of Eq. (3.22), and use the condition $\nabla \cdot \mathbf{B} = 0$, to arrive at an equation for the axial component of $\mathbf{B}$:

$$\nabla^2 B_z + \mu B_z = 0 .$$ (3.30)

The general solution of Eq. (3.30) is given by the real part of the expression

$$B_z = \sum_{m,k} a_{m,k} J_m(\alpha r) \exp\left[i(m\theta + kz)\right] ,$$ (3.31)

where $\alpha^2 = \mu^2 - k^2$, J_m is a Bessel function, and we have, for now, ignored the quantization $k = n/R$ introduced by the periodic boundary conditions. The r and θ components of $\mathbf{B}$ can then be found from the corresponding components of Eq. (3.22). It can be shown [*Taylor*, 1975; *Reiman*, 1980] that only two of the solutions given by Eq. (3.31) can have absolute minimum energy. These are: a cylindrically symmetric solution with $m = 0$

$$B_z / B_0 = J_0(\mu r) ,$$ (3.32a)

$$B_\theta / B_0 = J_1(\mu r) ,$$ (3.32b)

$$B_r / B_0 = 0 ;$$ (3.32c)

and, a helically symmetric state with $m = 0$ and $m = 1$

$$B_z / B_0 = J_0(\mu r) + a_{1,k} J_1(\alpha r) \cos(\theta + kz) ,$$ (3.33a)

$$B_\theta / B_0 = J_1(\mu r) + \frac{a_{1,k}}{\alpha} \left[\mu J_1'(\alpha r) + \frac{k}{\alpha r} J_1(\alpha r) \right] \cos(\theta + kz) ,$$ (3.33b)

$$B_r / B_0 = -\frac{a_{1,k}}{\alpha} \left[k J_1'(\alpha r) + \frac{\mu}{\alpha r} J_1(\alpha r) \right] \sin(\theta + kz) .$$ (3.33c)

Here, B_0 is the toroidal field at the magnetic axis, $r = 0$.

The symmetric states defined by Eqs. (3.32a-c) are known as the BFM. For these states, it can be shown [*Martin and Taylor*, 1974; *Taylor*, 1986] that the magnetic helicity K, the toroidal flux Φ_z, and the normalized parallel current density μa, are related by

$$\frac{K}{\Phi_z^2} = \frac{L}{2\pi a} \left\{ \frac{\mu a \left[J_0^2(\mu a) + J_1^2(\mu a) \right] - 2 J_0(\mu a) J_1(\mu a)}{J_1^2(\mu a)} \right\} , \qquad (3.34)$$

where L is the length of the cylinder. Thus, the details of the relaxed state are completely determined by the two invariants K and Φ_z: the ratio K/Φ_z^2 determines μ, and hence the field profiles; then, either K or Φ_z determines the field amplitude [Taylor, 1986]. It may be shown that the ratio K/Φ_z is proportional to the volt-seconds available from an external circuit used to drive the discharge [Taylor, 1974]. Thus, the profiles are related directly to experimentally controllable parameters. This is consistent with observations.

It is also interesting to interpret these results in terms of field reversal parameter $F = B_z(a)/\langle B_z \rangle = \pi a^2 B_z(a)/\Phi_z$, and the pinch parameter $\Theta = B_\theta(a)/\langle B_z \rangle = 2\pi a I_z/c\Phi_z$, where I_z is the total toroidal (axial) current, c is the speed of light, and $\langle .. \rangle$ denotes a volume average. Note that Θ is related to two fundamental global parameters characterizing the discharge: the total current and flux. For the BFM profiles, we find

$$F = \frac{\mu a J_0(\mu a)}{2 J_1(\mu a)} , \qquad (3.35)$$

$$\Theta = \frac{\mu a}{2} . \qquad (3.36)$$

Thus, the theory predicts $F < 0$ when $\mu a > 2.4$, or $\Theta > 1.2$. Whenever, for a given toroidal flux, enough volt-seconds are supplied to the plasma, the theory predicts that the toroidal (axial) field at the wall will change sign (or reverse) with respect to its value at $r = 0$. This is in substantial qualitative agreement with experiments. In Figure 3-2 we plot the theoretical F-Θ curve, as given by Eq. (3.35) and Eq. (3.36), along with experimental data. The curve represents the locus of minimum energy states.

We now return to the helically distorted state given by Eqs. (3.33a-c). This state exists only for discrete values of μ, which are determined by setting $B_r(a) = 0$. The lowest value of μ for which such solutions exist corresponds to $ka \approx 1.25$ ($\mu a \approx 3.11$, or $\Theta \approx 1.56$). The invariant K/Φ_z^2 no longer determines μ; it now determines the amplitude of the helical distortion $a_{1,k}$. When $\Theta > 1.56$, this is the state of lowest energy.

Thus, in Taylor's theory, for a given toroidal flux, as the volt-seconds are increased from zero, Θ will also increase. As Θ increases, the toroidal field at the wall will gradually diminish with respect to its value at the axis. As Θ exceeds 1.2, the field at the wall will reverse sign; Θ will continue to increase

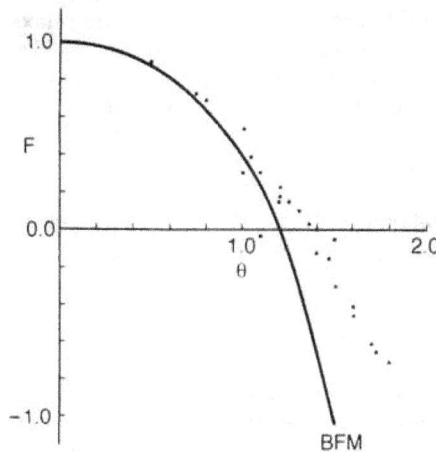

Figure 3-2. Taylor's F-Θ curve compared with experimental data.

(and F to decrease) until $\Theta = 1.56$. Then, as the volt-seconds are further increased, no further increases in Θ (current) will occur; rather, the increased volt-seconds will be absorbed inductively as the current channel deforms helically.

Summary of RFP Predictions

The predictions of Taylor's theory for the RFP can be summarized as follows:

1. The magnetic fields are characterized by uniform normalized parallel current density $\mu = J \cdot B / B^2$; they are thus independent of the initial conditions.

2. For $\Theta < 1.56$, the magnetic field profiles are described by the BFM Eqs. (3.32a-c).

3. For $\Theta > 1.2$, the axial field at the wall is reversed relative to its value at $r = 0$, i.e., F is negative.

4. The pinch parameter cannot be raised above $\Theta = 1.56$. If the volt-seconds are further increased, the relaxed state becomes helically distorted, as given by Eqs. (3.33a-c). The helical distortion increases the plasma inductance, hence absorbing the additional voltage without increasing the current. This helical state is characterized by a poloidal mode number $m = 1$, and by an axial mode number $ka \approx 1.25 > 0$.

5. Fully relaxed states have uniform pressure, i.e., they are force-free.

The Multipinch Experiment

Taylor's theory can be used to make predictions about other plasma configurations. For example, the multipinch is an axisymmetric toroidal discharge with a non-circular cross section: the poloidal cross section exhibits an equatorial, or up-down, symmetry as shown in Figure 3-3 [*LaHaye et al.*, 1986; *Taylor*, 1986; *Gimblett et al.*, 1987]. For such systems, the periodic cylindrical approximation used to describe the RFP does not apply, and toroidal effects must be included.

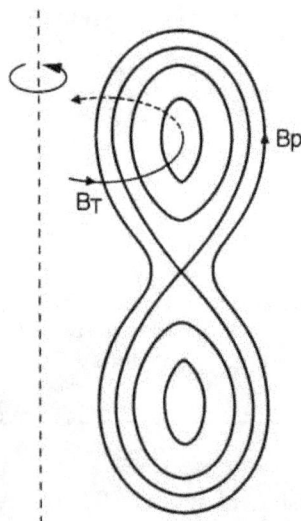

Figure 3-3. Sketch of multipinch configuration.

The calculation goes through in much the same way as for the RFP, except that Eq. (3.22) is now expressed as a partial differential equation in the poloidal plane; the details will not be given here. It turns out that physically interesting axisymmetric ($n = 0$) solutions can be found. The lowest energy state possesses equatorial (up-down) symmetry; in analogy with the RFP, this is the only solution for low values of K/Φ^2, where Φ is the toroidal flux. As in the RFP, the field profiles are parameterized by μa; K/Φ^2 determines μa, and then either K or Φ determines the amplitude. The eigenvalue problem analogous to Eq. (3.32c) and Eq. (3.33c) now determines states with equatorial asymmetry. The lowest eigenvalue for the equatorially asymmetric but still axisymmetric state (analogous to the helically deformed state in the RFP) corresponds to $\mu a = 2.21$. As the voltage is increased from a low value, μa increases until this value is reached. For larger voltage, μa remains fixed as the amplitude of the up-down asymmetry increases.

The predictions of the theory for the multipinch configuration are borne out by experiment [*LaHaye et al.*, 1986]. The current saturation at $\mu a = 2.21$, and the increasing up-down asymmetry for higher voltages, are all observed. Details such as the dependence of the saturation level on toroidal flux are also predicted with quantitative accuracy.

3.6 Discussion

We have discussed how Taylor's theory gives an accurate description of plasma relaxation in the case of the multipinch. While it gives a qualitative picture of relaxation in the RFP, it fails in some detailed predictions. These are briefly discussed here.

In the first place, the fully relaxed state with $\mu = constant$ is inconsistent with the boundary condition for a resistive plasma in contact with a rigid, perfectly conducting boundary. (Recall that resistivity does not explicitly enter the theory.) In that case, the current density must vanish at the plasma edge, i.e., $\mathbf{n} \times \mathbf{J} = 0$. This is seen by combining the resistive Ohm's law, Eq. (3.15), with the boundary conditions $\mathbf{n} \times \mathbf{E} = \mathbf{n} \cdot \mathbf{B} = \mathbf{n} \cdot \mathbf{v} = 0$. (This effect is amplified in real experimental plasmas with cold edges. The large edge resistivity then causes the current density to be small or zero near the wall.) In fact, real plasmas do not appear to be fully relaxed. This is illustrated in Figure 3-2; the experimental points consistently lie above the theoretically predicted F-Θ diagram, indicating incomplete relaxation. In Chapter 4 we will see that experimental measurements indicate a central region of relatively constant μ, with μ falling to zero at the wall.

The second point concerns plasma pressure. The assumptions made in the theory of completely relaxed states imply uniform pressure even when β

is finite. When combined with the physical boundary that the pressure must be small at the wall, this implies a state with $\beta << 1$. However, all RFP experiments exhibit $\beta \approx 0.1$.

A third point concerns the RFP current limit predicted to occur at $\Theta \approx 1.56$. Experimental RFP discharges do not tend to show a clear saturation in the maximum attainable Θ. Nonetheless, large amplitude, $m = 1$ fluctuations are associated with high Θ discharges. However, these characteristic fluctuations are observed to have $ka < 0$, so that they have *opposite* pitch from that predicted by the theory. One possible explanation for this discrepancy is that the helical state is excluded by the quantization introduced by periodicity of the z-coordinate. (This quantization is required if the z-coordinate is to approximate the periodic toroidal coordinate in a torus.) In that case, $k = n/R$, where n is an integer, so that the helically distorted solutions of Taylor can exist only for special systems with $R/a \approx n/1.25$. (Recall that the helical state has an eigenvalue $ka \approx 1.25$.) If the toroidal geometry in the experiment is not so constructed, these solutions are excluded.

In contrast, in the multipinch the corresponding eigenvalue problem distinguishes states of different equatorial, rather than axial, symmetry [*Taylor*, 1986]. These equatorially asymmetric states are not excluded by toroidal quantization (they all have $n = 0$). In this case, the predictions of the theory are quite accurate.

Finally, we draw an analogy between the theories of plasma relaxation and thermodynamics. Thermodynamics gives a set of laws that accurately describe the evolution of equilibrium systems as characterized by global parameters such as temperature, pressure, volume, etc.; these laws are useful as a calculus for computing the macroscopic equilibrium properties of such systems, but by themselves do not give a fundamental physical understanding. These laws were originally deduced empirically, with no reference to the dynamical processes occurring at the molecular level. For systems constrained to be at constant temperature, the proper macroscopic behavior is obtained by minimizing the Helmholtz free energy $F = W - TS$, where W is the internal energy, T is the temperature, and S is the entropy. Statistical mechanics gives us the connection between the microscopic, molecular picture and the macroscopic, thermodynamic picture by relating the partition function Z (which contains information about the laws governing the molecular motion) to the Helmholtz free energy (which contains the thermodynamic state variables) through the relationship $F = -T \ln Z$; only when this connection is made can the macroscopic behavior of thermodynamic systems be said to be well understood.

The status of the relaxation theory is similar, but incomplete. In Taylor's variational approach, we minimize a free energy function $I = W - \lambda K$, where λ is related to the normalized parallel current density and K is the magnetic helicity. For systems constrained to have constant helicity, the result is a useful calculus for computing the macroscopic equilibrium properties of magnetoplasma systems characterized by global variables such as current, helicity, and magnetic flux. It does not give an understanding of the detailed physical processes involved; it only requires that the underlying dynamics are describable in terms of resistive MHD. Unfortunately, at present there is no simple (or even complicated) formula, analogous to $F = -T \ln Z$, connecting the microscopic and macroscopic descriptions. In its place, we have a body of evidence, gained from experimental measurements and large scale numerical simulation of the resistive MHD equations, that addresses the relaxation process; the relevance of these results to physically observed plasmas determines the extent to which the relaxation process is well understood.

CHAPTER 4

PHENOMENOLOGY OF RELAXATION IN THE REVERSED-FIELD PINCH

In the previous chapters, we have presented a mathematical model appropriate to plasma relaxation, and a theory based on that model which describes the global properties of the final states of the relaxation process. A further quantitative description of real plasmas requires that either detailed solutions of the nonlinear MHD equations be obtained, or that simpler models which incorporate the essential physics be developed. Detailed solutions are described in Chapter 5; here we present a phenomenological model which has proven useful in describing the real experimental operation of RFP plasmas. Many of the results of the nonlinear MHD numerical simulations found in Chapter 5 can be predicted on the basis of this relatively simple picture.

The phenomenological model presented is instructive for several reasons: it can be conceived in principle without knowledge of Taylor's theory of fully relaxed states, it does not require a detailed knowledge of nonlinear MHD dynamics, and it gives a fairly accurate description of many experimental observations presented and discussed in the last part of this chapter. Specifically, we first develop simple models of the mean magnetic field profiles as deduced from experimental observations. We then apply basic linear resistive MHD stability analysis, as discussed in Chapter 2, to determine the stability properties of these profiles. Finally, we analyze the effect of resistive diffusion on these profiles.

The phenomenological model of the RFP will emerge as a combination of the above separate issues. For simplicity, we will consider them as independent phenomena; in reality, they operate in concert.

In particular, we will show that linear MHD stability theory can provide a very useful and relatively simple tool to interpret experimental results. By characterizing the measured mean field profiles with simple parametric analytic expressions, a systematic stability analysis can be performed. As a result, we will find that MHD stability sets clear limits on both the profile shape, and on the magnitude of the current density on axis. These limits correspond very well to those found experimentally, as well as with the results of nonlinear MHD numerical simulations. We will also show that resistive diffusion is always a destabilizing process, and that an initially stable profile can be driven across an MHD stability boundary in a time that is a small fraction of the time required for resistive diffusion of the global profile.

4.1 Mean Field Profiles

The study of mean field profiles generically similar to those found in the RFP has a long history, ranging from the field of controlled thermonuclear fusion research [*Butt et al.*, 1958; *Colgate et al.*, 1958; *Ohkawa*, 1963], MHD stability theory [*Rosenbluth*, 1958; *Robinson*, 1971; *Freidberg*, 1987], mean field electrodynamics [*Moffatt*, 1978; *Krause and Rädler*, 1980], to astrophysical plasmas [*Woltjer*, 1958; *Heyvaerts and Priest*, 1984; *Könial and Choudhuri*, 1985]. These various approaches find a unifying thread in Taylor's theory of relaxed states [*Taylor*, 1974] where the ideal MHD result of *Woltjer* [1958] is extended to dissipative fluids, and the invariance of the total helicity is used to determine the minimum energy magnetic distribution. This theory is developed in Chapter 3. In particular, for a plasma surrounded by a perfect flux conserving shell the minimum energy states are those satisfying the force-free equation

$$\nabla \times \mathbf{B} \;=\; \mu \mathbf{B} \tag{4.1}$$

with $\mu = constant$.

As described in Chapter 3, for values of $\mu a = 2\Theta < 3.11$ $\left(\text{where } a \text{ is the plasma radius and } \Theta = B_\theta(a)/\langle B_z\rangle\right)$, the minimum energy states are described by the cylindrical solution given by Bessel functions as $B_\theta = B_0 J_1(\mu r)$, and $B_z = B_0 J_0(\mu r)$, and is often denoted as the BFM.

This result can account for many experimental observations and also proves in a general way the MHD stability of force-free fields with constant μ. (Further instability is impossible since these states already possess minimum energy subject to the relevant constraints.) The experimental profiles deviate from this simple theoretical prediction due to the presence of pressure gradients; in particular, to temperature gradients related to the attempt to achieve hot plasma confinement away from the cold edge near the wall. An important point is that the finite pressure gradient plays a twofold role: (1) it introduces a perpendicular current density component that, although small, contributes to the total current density that determines the magnetic field; and (2) it corresponds to a resistivity gradient that enhances the relative dissipation in the outer region. Relaxation must therefore act more strongly in this region to oppose large departures from the minimum energy state. The overall result is that, although the profile is fairly close to the fully relaxed $\mu = constant$ state in the central plasma region where the pressure gradient is small, in the lower temperature, higher resistivity outer region generally less current can be maintained than that corresponding to the fully relaxed profile.

To construct more realistic profiles and to separate the contribution of the parallel and perpendicular components of current density, a model can be used in which the variations in the magnetic field profiles due to the plasma pressure gradient (which are usually small, with $\beta \sim 0.1$) are separated from those determined by the larger force-free parallel current component [Ortolani, 1984]. Also, the interpretation of the MHD stability properties of the equilibrium profiles is clarified by this description which separates the instability driving force related to the parallel current from that related to the pressure. The force free fully relaxed case is recovered in the case $\nabla p = 0$ and $\mu = constant$.

The nondimensional equations of the model in cylindrical geometry are listed below:

$$\frac{\beta_0}{2} \, \nabla p \; = \; J \times B \tag{4.2}$$

$$\nabla \times B \; = \; J \; = \; J_{\parallel} + J_{\perp} \tag{4.3}$$

where

$$J_{\parallel} \; = \; \frac{J \cdot B}{B^2} \, B \; = \; 2\Theta_0 \, \mu B \tag{4.4}$$

$$J_{\perp} \; = \; -\frac{J \times B}{B^2} \times B \; = \; -\frac{\beta_0}{2} \, \frac{\nabla p \times B}{B^2} \tag{4.5}$$

and

$$\Theta_0 = \frac{\mu(0)a}{2}, \; \beta_0 = \frac{8\pi p(0)}{B_0^2} \; . \tag{4.6}$$

Then *specifying* $\mu(r)$ and $\nabla p(r)$ determines $B(r)$ as:

$$\frac{\partial B_z}{\partial r} \; = \; -2\Theta_0 \, \mu B_\theta - \frac{\beta_0}{4} \, \frac{\partial p}{\partial r} \, \frac{B_z}{B^2} \tag{4.7}$$

$$\frac{\partial B_\theta}{\partial r} \; = \; -\frac{B_\theta}{r} + 2\Theta_0 \, \mu B_z - \frac{\beta_0}{4} \, \frac{\partial p}{\partial r} \, \frac{B_\theta}{B^2} \; . \tag{4.8}$$

Note that for the BFM, $\Theta_0 = \Theta$. In general, by specifying Θ_0 and β_0 and for given profiles of μ and p, the field profiles B_z and B_θ (and the corresponding J_z, J_θ or $J_{||} J_\perp$), are easily found integrating Eq. (4.7) and (4.8).

An example of the profiles obtained with $\Theta_0 = 1.6$, $\beta_0 = 0$, and $\mu = 1 - r^4$ is given in Figure 4-1, where the BFM is also drawn for comparison. It is seen that by considering a μ profile that is fairly constant in the inner region, but decreases in the outer region to match the $\mu(a) \approx 0$ condition, the unrealistic features of the BFM are removed. Considering the effect of finite β (i.e., $J_\perp \neq 0$) slightly affects the profiles. An example is shown in Figure 4-2 where $\beta_0 = 0.1$ and parabolic density and temperature profiles have been assumed to describe the pressure $p = 2nT$.

Experimental results are well represented by field profiles of this sort. Examples of measured profiles (solid lines) and fully relaxed profiles (dashed lines) corresponding to $\beta = 0$ and $\mu = $ const. are shown in Figure 4-3.

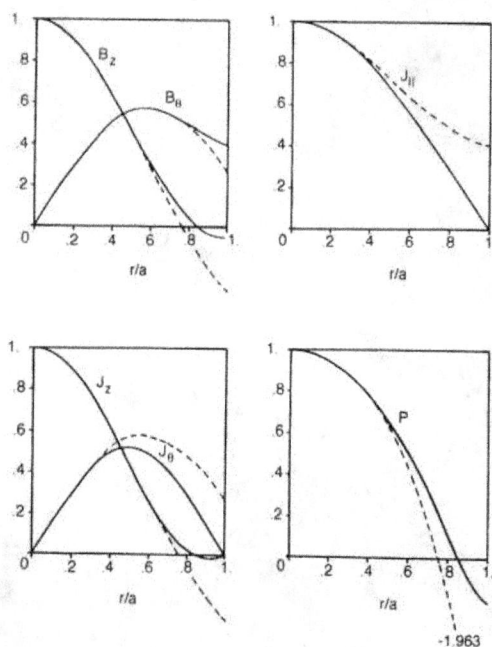

Figure 4-1. Profiles corresponding to $\mu = 1 - r^4$, $\Theta_0 = 1.6$, $\beta_0 = 0$. The BFM profiles are shown for comparison (dashed lines).

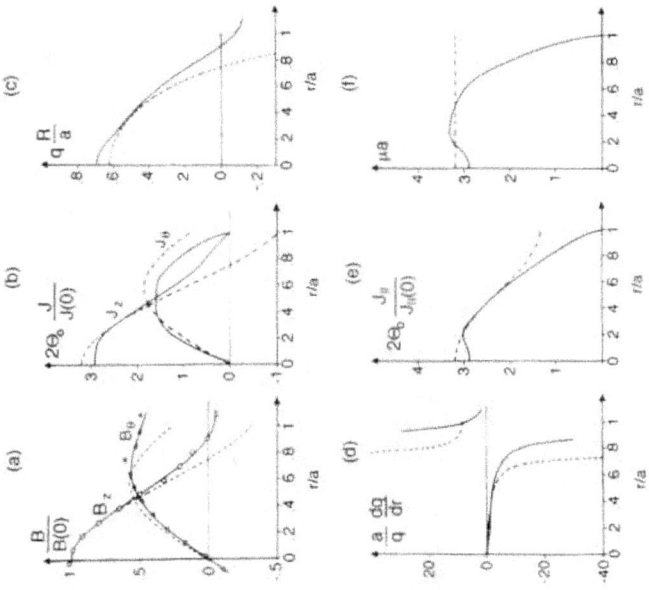

Figure 4-3. Radial profiles of **a.** magnetic field **b.** current density **c.** safety factor **d.** shear **e.** parallel current density, and **f.** μ.

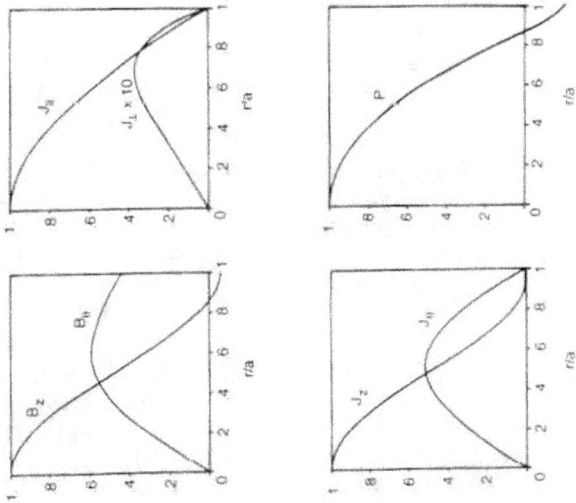

Figure 4-2. Profiles corresponding to $\mu = 1 - r^4$, $\Theta_0 = 1.6$, $\beta_0 = 0.1$.

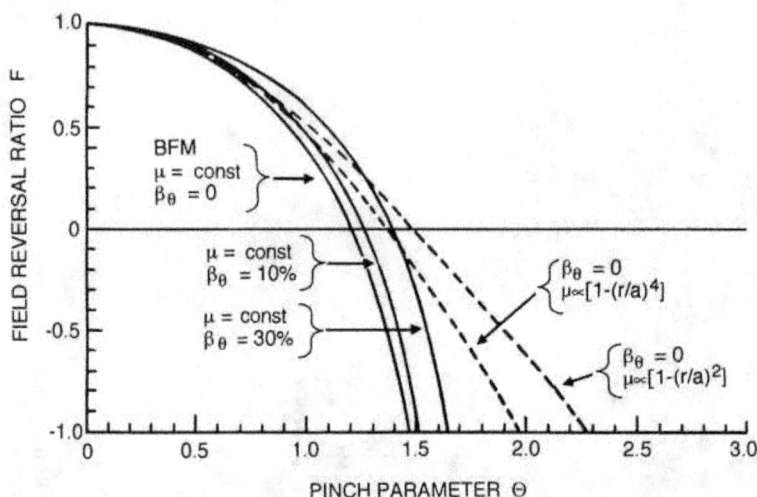

Figure 4-4. Calculated $F-\Theta$ curves with magnetic field profiles generated by assuming that μ varies with radius as $\mu(r) = (2\Theta_0/a)\,(1-(r/a)^\alpha)$. The effect of changing the radial field distribution by choosing different values of α, and of varying β_θ up to 30% is shown. For comparison, the $F-\Theta$ curve for the BFM with $\mu(r) = constant$ and $\beta_\theta = 0$ is also plotted.

Diagrams of the $F-\Theta$ dependence are shown in Figure 4-4 for the BFM and for equilibria described by

$$\mu = \frac{2\Theta_0}{a}\left[1-(r/a)^\alpha\right] \tag{4.9}$$

Profiles for $\mu(r) = constant$ with various values of β_θ ($= 8\pi p/B_\theta^2$) and different $\mu(r)$ profiles with $\beta_\theta = 0$ are plotted. In general, with values of $\beta_\theta \leq 30\%$, the effect of a parallel current profile with $\mu(r) \neq constant$ is dominant compared to the perpendicular current contribution in displacing the $F-\Theta$ curve to the right of that for the BFM. Note that values of Θ well above the theoretical maximum of 1.56 can be obtained.

In attempting to explain why such large values of Θ are observed, we note that the usual definition of Θ is heavily weighted towards the outer regions where the profiles differ most from those of BFM. The parameter $\Theta = \mu a/2$ in the BFM can be redefined [Ortolani, 1984] in terms of μ or q on the axis as $\Theta_0 = (a/R)\,(1/q(0)) = \mu(0)a/2$.

In Figure 4-5, the parameter Θ_0 is plotted as a function of Θ for the BFM and for the model based on Eq. (4.9). Experimental data fit to Eq. (4.9) is also shown [*Ortolani*, 1985]. It is seen that Θ_0 does tend to saturate, at a value of about 1.8, although Θ is observed to increase up to 2.5 or more. The limit in Θ_0 corresponds to a limit on the current density in the central regions, whereas according to Taylor's theory the Θ-limit corresponds to a limit in the *total* current. The limit in Θ_0, corresponding to a lower limit of the values of q on axis, $q_0 \gtrsim (2/3)(a/R)$, is similar to the $q_0 > 1$ limit in tokamak, although the details are different. Dynamical details of the q_0 limiting process are described in Section 4.4, and in Chapter 5.

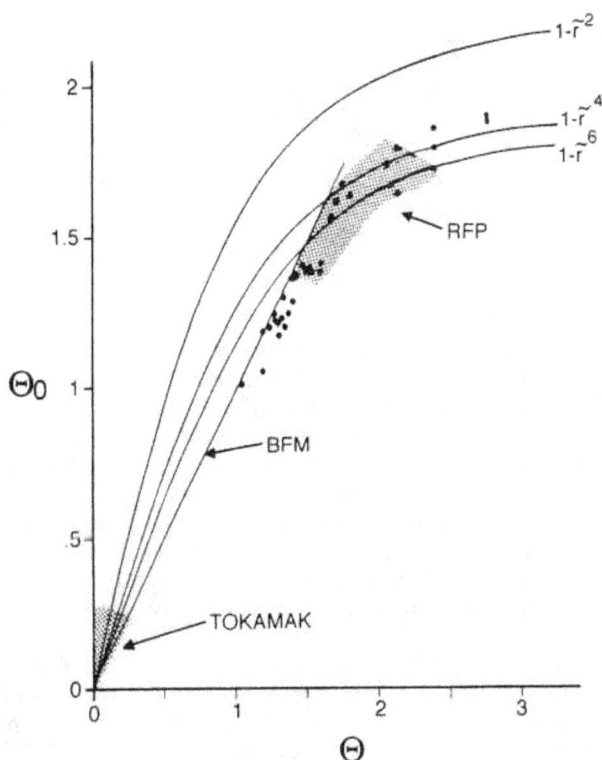

Figure 4-5. Θ_0 vs. Θ diagram showing the operating region for a tokamak and an RFP, experimental data and various curves corresponding to different α values.

The two typical regions of operation for tokamaks and RFPs are shown shaded in the above figure. Although the minimum energy state theory may also apply to tokamaks provided toroidal effects are taken into account [*Taylor*, 1984] it has not proven so fruitful in that case. As discussed in Chapter 1, the tokamak operates further away from the Taylor relaxed state than the RFP, and when it does relax the process is more violent [*Kadomtsev*, 1977]. The tokamak possesses stability because of periodicity and magnetic well effects. One can speculate that because in the RFP with $q < 1$ there are many singular surfaces, unstable modes will always exist that allow continuous relaxation and therefore relatively little departure from a relaxed state.

4.2 The Stability of Relaxed States

The properties of relaxed states can be discussed either in terms of relaxation theory or stability theory, and the approaches are complementary. The true minimum energy relaxed state is stable to all modes as there is no free energy available to drive instabilities. In practice we are interested in states which differ to some extent from the BFM and therefore have higher energy. These states are subject to the MHD instabilities that are responsible for relaxation.

The linear MHD stability analysis is relatively simple but very instructive, for it determines general constraints that must be satisfied by the mean magnetic field profiles in order to maintain MHD stability. Of course, the theory contains the fully relaxed minimum energy state, but it also provides clear stability criteria for more realistic near-minimum energy profiles. In particular, the analysis shows the existence of a limit on q_0, the value of q on the magnetic axis, which is in good agreement with experimental results. Indeed, the q_0 limit, plus constraints on the profile shapes that also emerge from the analysis presented here, allow a realistic description of the mean field profiles about which quasi-periodic oscillations occur in both experiments and in nonlinear numerical simulations.

Various models have been introduced to account for magnetic field profiles that are more realistic than the BFM, and the corresponding MHD stability properties have been discussed by many authors. Some stability criteria have been deduced [*Bodin and Newton*, 1980]. An ideal MHD stability analysis has been carried out [*Robinson*, 1971] for diffuse pinch distributions. RFPs are found to be stable if the pitch of the magnetic field lines $P(r) = rB_z/B_\theta = Rq(r)$ is a monotonically decreasing function of radius, provided the total toroidal flux is positive and $P(a) > -3P(0)$. In a later paper [*Robinson*, 1978], it is shown that ideally stable configurations can also be made tearing mode stable by moving a perfectly conducting wall surrounding the plasma closer to the plasma. However, in these models Θ can be varied simply by changing the position of the conducting wall surrounding the plasma. The field profiles near the magnetic axis do not depend on Θ. We shall see that these limitations do not allow for the assessment of the important modes that are resonant in the core of the RFP.

On the other hand, we have seen in Section 4.1 that the RFP configuration can be described by a simple expression of the ratio of parallel current density to magnetic field as

$$\mu = \frac{J_{\parallel}}{B} = \mu\,(0)f(r)\,, \tag{4.10}$$

where $\mu(0) = 2/Rq(0) = 2\Theta_0/a$ characterizes the on-axis value of the current density profile. The form function $f(r)$ is assumed given by: $f(r) = 1 - (r/a)^\alpha$, with α representing the width of the current channel. In this model, two parameters, Θ_0 and α, completely describe the force-free part of the profile, and a parametric MHD stability analysis can be performed [Antoni et al., 1986]. We review here the results of this linear resistive MHD stability analysis which provides a basis for interpreting the dynamics of partially relaxed states.

Various RFPs equilibrium profiles measured in different experimental conditions can be reproduced by varying α in the range $2 \le \alpha \le 6$ and Θ_0 in the range $1.4 \le \Theta_0 \le 2$.

The stability with respect to ideal modes can be analyzed by solving the linearized MHD equations in cylindrical geometry as an eigenvalue problem, so that the radial dependence of the perturbation and the growth rate can be determined. For the linear *resistive* stability analysis, the results presented here are based on the Δ' *criterion* [Furth et al., 1963], where Δ' is the discontinuity in the logarithmic derivative of the perturbed radial magnetic field at the resonant surface. As described briefly in Chapter 2, Δ' is determined for given mode numbers m and k by solving the linearized ideal MHD equations away from the rational surface $r_{m,k}$ (where $\mathbf{k} \cdot \mathbf{B} = 0$) subject to the proper boundary conditions at $r = 0$ and $r = a$. Resistive instability results if $\Delta' > 0$. For each equilibrium configuration associated with given values of α and Θ_0, we can consider the stability for a wide spectrum of Fourier modes identified by toroidal wavenumber k and poloidal mode number m. In particular, $m = 0,1,2$ are considered, and the toroidal wavenumber is normalized as $k = 2\pi a/\lambda$.

In the range of α and Θ_0 defined above, the $m = 1$ resistive modes are found to impose the more severe stability boundaries. Ideal $m = 0$ modes are completely stable, while $m = 0$ resistive modes arise in configurations where $m = 1$ unstable tearing modes are already present. Ideal and resistive $m = 2$ modes are always found to be stable.

To illustrate how the stability properties depend on the parameters characterizing the current density profile, we first discuss some examples of the results obtained for the $m = 1$ modes. We see that the most important modes are $m = 1$ resistive modes which are resonant between the axis ($r = 0$) and the field reversal surface. These modes are found because of the characterization (Eq. 4.9) used for the profiles. They would not appear if $\mu = constant$ (the BFM), nor for the class of profiles defined by Robinson [*Robinson*, 1978], where a variation of Θ corresponds only to a change in the wall position and not to a change in the actual profile shape or magnitude near the magnetic axis. However, in order to compare with previous stability analysis of the BFM [*Voslamber and Callabaut*, 1962; *Gibson and Whiteman*, 1968; *Robinson*, 1971, 1978], we begin with a fairly flat μ profile with $\alpha = 6$ and a relatively high current density on axis, $\Theta_0 = 2$.

For this profile, we obtain the configuration shown in Figure 4-6a, which is characterized by a negative total toroidal flux and a deep toroidal field reversal. These features lead to the loss of the stabilizing effect of the conducting wall. In this case we expect, from Robinson's theory [*Robinson*, 1971, 1978], to find unstable external modes, i.e., modes with $k > 0$ that are resonant outside the reversal surface. Indeed, for the case considered the analysis shows the existence of $m = 1$ modes, both ideal and resistive, in the range $0.3 \leq k \leq 1.1$ and $1.1 \leq k \leq 3$, respectively. On the other hand, internal modes, i.e. modes with $k < 0$ that are resonant inside the reversal surface, are found to be stable, because of the relatively flat μ profile in the inner region. Keeping the values of α constant at $\alpha = 6$ and decreasing Θ_0 to $\Theta_0 = 1.8$ (Figure 4-6b), we find that the external modes are still unstable, but the range of unstable wavenumber, $0.9 \leq k \leq 1.9$ is smaller than in the previous case, owing to the closer wall position, as can be seen in the figure.

If Θ_0 is decreased further to $\Theta_0 = 1.6$ (Figure 4-6c), the configuration becomes completely stable. In this case, the fairly flat μ profile in the central region prevents the internal modes from developing and the conducting wall stabilizes the external modes. Decreasing α to $\alpha = 4$, we find external unstable modes for $\Theta_0 = 2$, but in this case the RFP configurations become completely stable for $\Theta_0 \leq 1.9$.

Now consider a peaked current density profile with $\alpha = 2$ and $\Theta_0 = 2$ (Figure 4-6d). We find that the external modes are stable but internal modes become unstable. In particular, in this case we find ideal and resistive $m = 1$ unstable modes in the ranges $-2.1 \leq k \leq -1.8$ and $-3.1 \leq k \leq -2.1$, respectively.

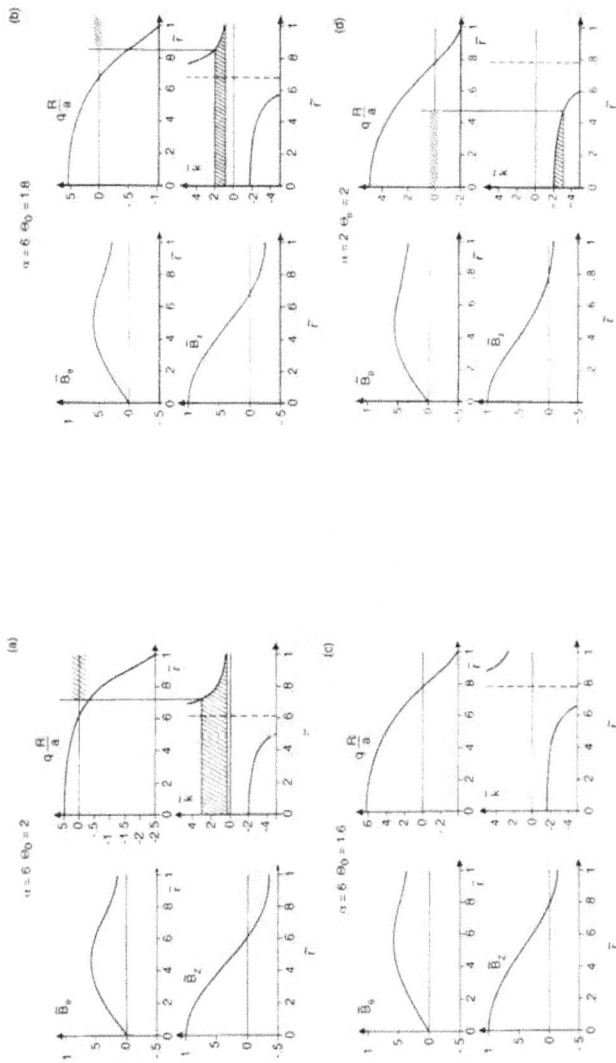

Figure 4-6 . Radial profiles of B_θ, B_z, qR/a and resonant k ($m = 1$) for RFP configurations with a. $\alpha = 6$, $\Theta_0 = 2$; b. $\alpha = 6$, $\Theta_0 = 1.8$; c. $\alpha = 6$, $\Theta_0 = 1.6$; d. $\alpha = 2$, $\Theta_0 = 2$. Dashed region in k diagram indicates unstable k range; associated resonance radial region is represented in qR/a diagram.

If we increase α up to $\alpha \geq 2.7$, the profile becomes completely stable. If Θ_0 decreases, the minimum α for stability with respect to internal modes increases up to $\alpha = 3.9$ for $F = 0$.

Figure 4-7 shows some example of the radial component of the magnetic field perturbation for typical unstable modes, showing their fairly large radial extent.

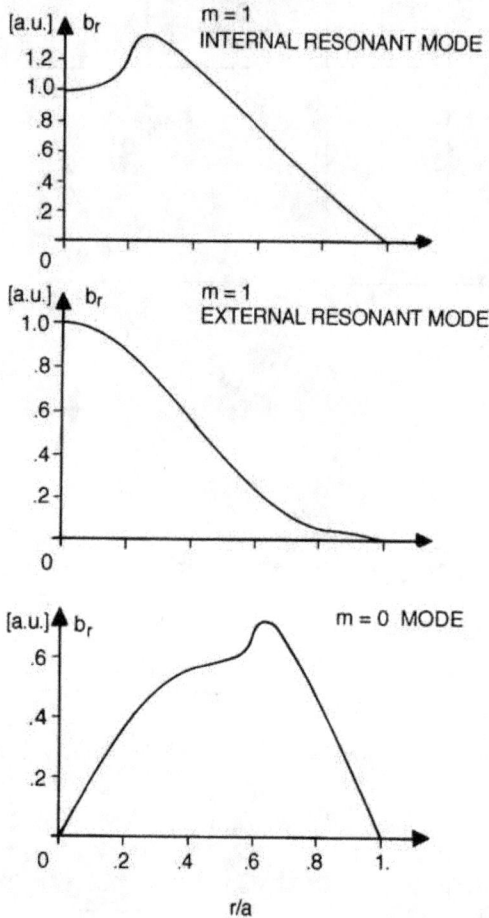

Figure 4-7. Typical profiles of radial component of magnetic field perturbation for $m = 1$ ($k = -1.8$ and $k = 1.6$) and $m = 0$ ($k = 0.2$) unstable modes.

In Figure 4-8, we have summarized the results of the ideal and resistive MHD analysis for the $m = 1$ case in the $\alpha - \Theta_0$ plane. The region of RFP existence is bounded by the curve $F = 0$, which corresponds to the transition between a stabilized pinch without toroidal field reversal and an RFP, and by the curve $F = \infty$, which corresponds to vanishing toroidal flux. Inside this region three other curves are reported which bound the zone of complete MHD stability for current driven modes.

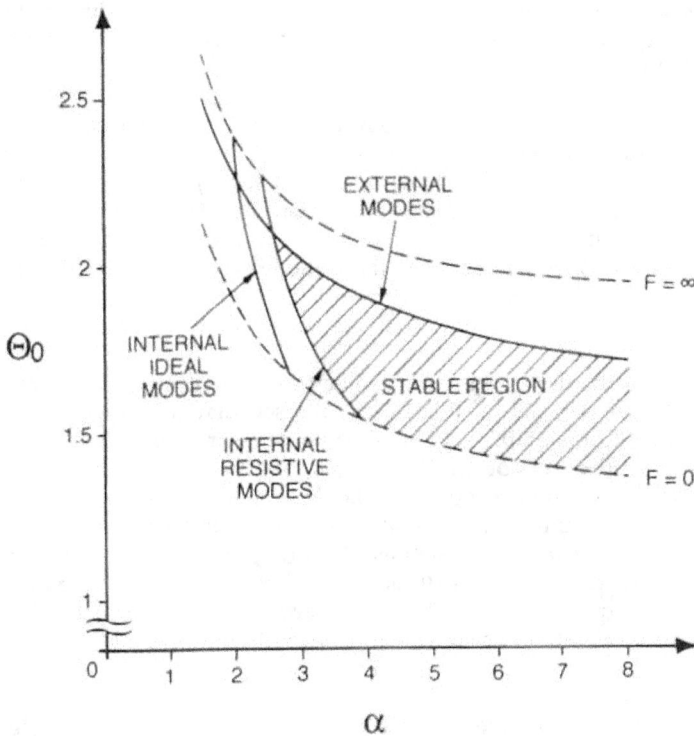

Figure 4-8. Stability diagram for $m = 1$ modes in the α–Θ_0 plane.

The upper curve limits the region where both ideal and resistive modes resonant outside the field reversal are unstable. These modes are analogous to those found for the BFM [*Gibson and Whiteman*, 1968] when Θ exceeds 1.56, and are due to the inward displacement of the reversal surface and decrease of the toroidal flux, ultimately leading to the loss of wall stabilization. In practice, these modes are rarely seen in experiments; they can easily be avoided by reducing the value of Θ. Furthermore, we remark again that in the BFM (μ = *constant*), or in the description used by Robinson [*Robinson*, 1978], Θ can be varied by changing the position of the conducting wall; the field profiles near the magnetic axis do not depend on Θ. This is an unrealistic feature and, in general, it is found that the form of the profile is related to the on-axis value of Θ [*Antoni et al.*, 1989]. Therefore, a description with at least two parameters, Θ_0 and α, is essential and highlights the presence of another stability boundary, that for internal modes: helical perturbations resonant between the axis and field reversal surface.

Figure 4-8 also shows the boundary curve for internal modes. For $\alpha < 3.9$, *internal* resistive modes are found, and for $\alpha < 2.8$, the spectrum of these instabilities also contains ideal modes, including the mode resonant on-axis that characterizes these fairly peaked μ profile. These modes do not have a BFM analogy, since in the BFM the driving term due to the gradient in the μ profile vanishes. We shall see that these modes with $m = 1$ are in fact those seen in experiments and found in nonlinear MHD numerical simulations. They are the modes that govern the dynamics of the RFP profile.

The $m=1$ stability results can also be represented in the plane $\Theta - \Theta_0$, as shown in Figure 4-9. From this figure, we see that the range of Θ values corresponding to the stable region is rather wide, ranging from ~1.5 to ~ 4, while the related excursion in Θ_0 is much less, between ~ 1.5 and ~ 2. The dashed lines in the figure represent the Θ_0 versus Θ values obtained with $\alpha = 2, 4, 8$. From the figure, we also observe that only with a rather flat parallel current distribution ($\alpha \sim 8$) would it be possible, for realistic values of $\Theta_0(< 2.5)$, to get closer to the external mode boundary. Indeed, as discussed later, typical experimental RFP profiles are normally quite far from this boundary, but operate close to the limit posed by the internal $m = 1$ resistive modes.

In summary, from the analysis, we find that a completely stable region exists for ideal and resistive current driven modes. In particular, for the stability of internal resistive and ideal $m = 1$ modes, minimum values of α have been found, as $\alpha = 3.9$ and $\alpha = 2.8$, respectively. The stability boundaries found can also be drawn (see Figure 4-10) in terms of the on-axis q, $q(0)$, and q at the wall, $q(a)$. The same figure also shows the ranges of experimental values of $q(0)$ and $q(a)$ for typical RFP discharges in different operating

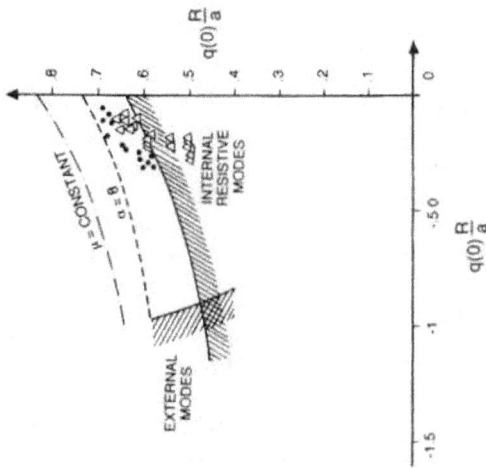

Figure 4-10. Stability diagram for $m = 1$ modes in $q(a)R/a$–$q(0)R/a$ plane; dots and triangles indicate, respectively low Θ and high Θ experimental configurations. Also shown are curves for fully relaxed μ = constant states with Θ_0 in range 1.2–1.56 and curve for our model with $\alpha = 8$.

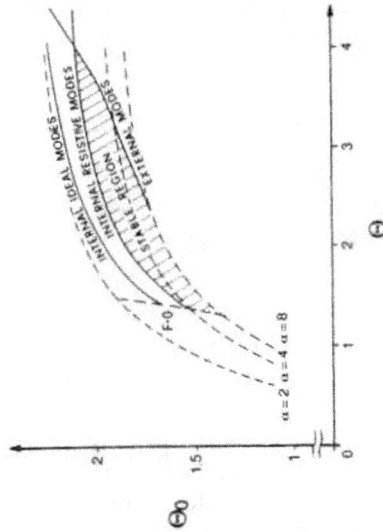

Figure 4-9. Stability diagram for $m = 1$ modes in Θ–Θ_0 plane.

conditions. The dots indicate relatively low Θ discharges, while the triangles represent relatively high Θ, sustained discharges.

The mean field profile for low Θ RFPs lies in the stable region below the uniform μ curve corresponding to Taylor's fully relaxed states and, as was remarked previously, is close to the $m = 1$ resistive internal modes boundary. However, high Θ discharges typically lie around this boundary. In this way, it is found that the external modes are not very significant in that they arise in conditions which are only rarely produced experimentally. On the other hand, a lower limit on q_0 can be identified as $q_0 > 2a/3R$. This limit is slightly dependent on the μ profile (varying between $\sim 2a/3R$ for $\alpha = 3.9$ and $\sim a/2R$ for $\alpha = 2.7$) and is associated with $m = 1$ internal resistive unstable modes. This result confirms the existence of such a limit as previously derived from experimental measurements of the mean field profiles [Ortolani, 1984]. This limit ultimately imposes an upper limit on the on-axis current density that can be maintained with a given magnetic field. In a non-dimensional form, this is given by $J(0) < 3B(0)$. Thus, these instabilities can provide an interpretation both for the upper limit at which the saturation of Θ_0 with increasing Θ is experimentally observed and for the experimental oscillations observed around the mean field profiles. These oscillations are particularly evident in high Θ discharges as discussed in the next sections.

4.3 Resistive Diffusion

If the RFP were dominated by classical diffusion the lifetime of the configuration would be severely limited. The ordinary diffusion equation for a static conductor of uniform nondimensional resistivity η,

$$\frac{\partial \mathbf{B}}{\partial t} = \eta \nabla^2 \mathbf{B} , \tag{4.11}$$

has the cylindrically symmetric solution

$$B_\theta = B_0 J_1 (\mu r) e^{-t/\tau} , \tag{4.12}$$

$$B_z = B_0 J_0 (\mu r) e^{-t/\tau} , \tag{4.13}$$

with $\tau = 1/\eta\mu^2 = a^2/4\eta\Theta^2$, which with $\Theta_0 \approx \Theta$ gives $\tau \approx (1/\eta)\left(P^2(0)/4\right)$.

The MHD stability of the BFM requires $\mu a < 3.176$ for ideal modes [Voslamber and Callabaut, 1962], and $\mu a < 3.104$ for $m = 1$ resistive modes [Gibson and Whiteman, 1968]. The latter corresponds to the $\Theta < 1.56$ limit in Taylor's theory. Then we have approximately $\tau = a^2/(10\eta)$. However, the

BFM profile with $\nabla p = 0$ preserves its radial structure while it decays, and the stability properties remain unchanged in time. Therefore, the previously estimated times represent an upper limit compared to more realistic calculations of the resistive diffusion of nonuniform μ and η profiles.

On the other hand, experimentally it is observed that a discharge with $\Theta = constant$ can be maintained for times significantly longer than the diffusion time. In particular, it is found that when $\langle B_z \rangle$ is fixed the reversed field configuration is sustained as a quasi-stationary state as long as the toroidal current lasts. The magnetic field is apparently generated at a rate which just compensates the tendency of the configuration to decay by field diffusion.

For example, in Figure 4-11 [*Caramana and Baker*, 1984] results are shown where the toroidal plasma current and the toroidal magnetic field at the wall were held constant by the external circuits. It is seen that the experimentally measured toroidal flux also remains approximately constant for about 20 ms, compared with a decay time of about 5 ms calculated using a resistive diffusion model with isotropic classical resistivity and no internal field generation process.

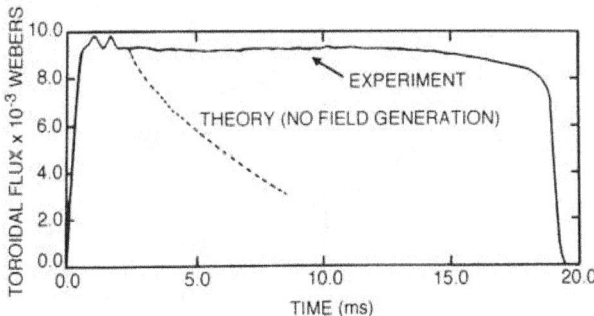

Figure 4-11. Toroidal flux vs. time with total current and magnetic field at the wall held constant. The measured time variation is compared with that calculated assuming classical field diffusion without field generation [*Caramana and Baker*, 1984].

Theoretically, a stationary RFP configuration is not possible even with the assumption of an anisotropic resistivity. In fact, the poloidal component of Ohm's law for a cylindrical geometry can be written as:

$$E_\theta - v_r B_z = \eta_\| J_\theta \left[1 + \left(\frac{\eta_\perp}{\eta_\|} - 1 \right) \frac{B_z^2}{B^2} \right] - J_z B_\theta \frac{B_z}{B^2} \left[\eta_\perp - \eta_\| \right] \tag{4.14}$$

which, for stationary conditions that correspond to profiles with $E_z = constant$ and $E_\theta = 0$, gives $J_\theta = 0$ where $B_z = 0$. We can now consider two practical cases: the toroidal (axial) field at the outer boundary is maintained at a constant negative value by controlling the current in the toroidal field coils; and, the total toroidal magnetic flux is held constant by applying zero poloidal voltage. In the first case, the RFP will decay to a state with B_z negative everywhere (see Figure 4-12a); in the second case, field reversal will be lost (see Figure 4-12b). The actual time depends on the resistivity.

The limit $\eta_\| / \eta_\perp = 0$ gives a static equilibrium which is known as the force free paramagnetic model (FFPM) [Rusbridge, 1962; Burton et al., 1962]. In this case a static solution shown in Figure 4-13 is possible with $B_z(a) = 0$ but not reversed. We thus see that in order to maintain a steady-state RFP profile some mechanism is necessary that regenerates the magnetic field in the presence of ordinary field diffusion.

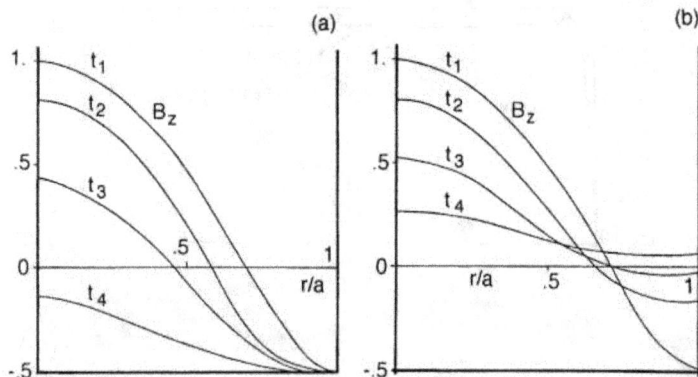

Figure 4-12. Examples of the time evolution of B_z for an RFP configuration corresponding to the magnetic field diffusion with boundary conditions imposing a. constant field at the wall or b. constant flux.

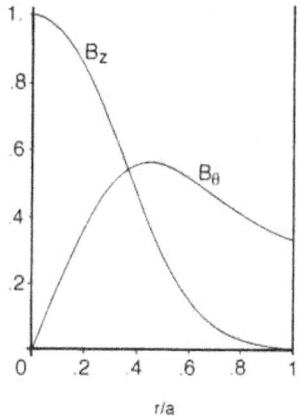

Figure 4-13. Stationary distribution of the magnetic field components for the force-free paramagnetic model.

However, although experimentally resistive diffusion does not limit the global time duration of the magnetic configuration, it has an important role in determining the observed dynamics. In fact, although *on average* the experimental magnetic field profiles are stable to current driven modes, resistive diffusion tends to destabilize the configuration by peaking the current distribution on axis and therefore leading to lower $q(0)$. A particular example, often realized in experiments, is that of sustained discharges where both the plasma current and the toroidal magnetic flux are constant in time. With these boundary conditions, resistive diffusion would lead to a decrease in $q(0)$. When $q(0)$ drops below a critical value, $q(0) \approx (2/3)(a/R)$, internal current driven instabilities can be excited as discussed in Section 4.2. This process is described in detail in Chapter 5.

For example, if we choose as an initial condition a stable configuration, we can approximately follow its temporal evolution due purely to resistive diffusion in the $\alpha - \Theta_0$ plane. Examples of typical trajectories of diffusing profiles are drawn in Figure 4-14. One can see that the time evolution, characterized by a constant value of Θ, tends to decrease α and to increase Θ_0, thus peaking the current distribution on axis. An analogous behavior is found if the boundary condition of constant toroidal field at the wall is used instead of constant toroidal flux. It is important to note that the transitions shown in Figure 4-14 from a stable to an unstable profile through the $m = 1$ internal mode boundary take place in a small fraction of the global profile resistive diffusion time, typically $1/10 - 1/100$.

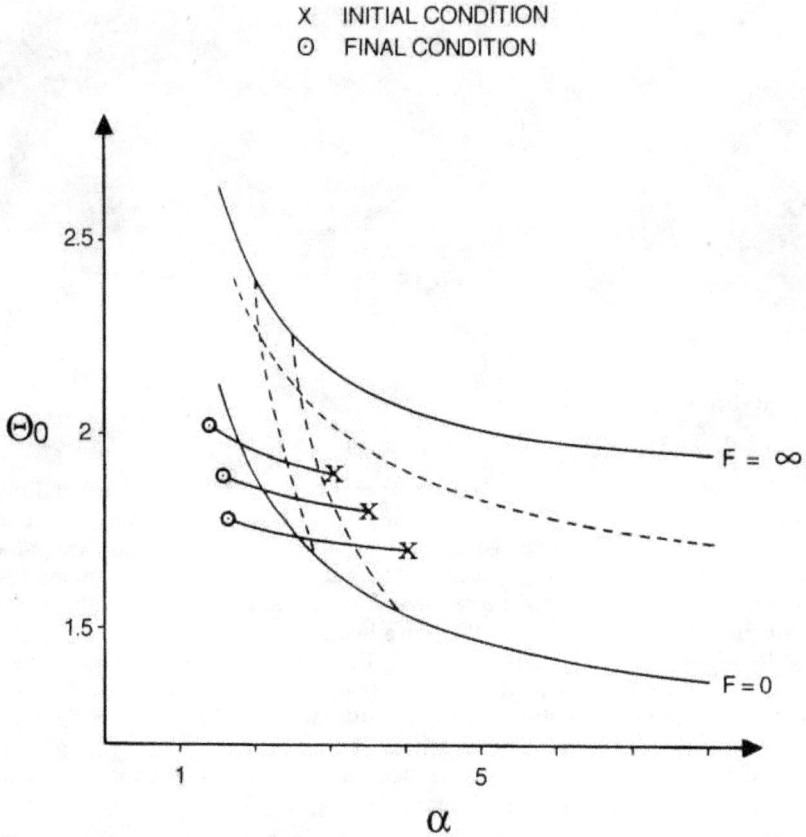

Figure 4-14. Trajectories in the $\alpha - \Theta_0$ plane due to resistive diffusion.

4.4 The Phenomenological Cyclical Model

The above results and observations suggest a description of relaxation phenomenon and RFP dynamics in terms of a cyclical model [Ortolani, 1984, 1987]. This concept is based on the observation that, whereas on the one hand the relaxation process draws the configuration towards the $\beta = 0$, $\mu = constant$ state, thereby disposing of excess magnetic and kinetic energies, on the other hand, magnetic field diffusion and its associated plasma heating are mechanisms of departure from the minimum energy state, which increase the deviation from $\mu = constant$ and $\beta = 0$. Due to resistive diffusion, the on-axis value of q drops as the current density profile peaks and shrinks.

As seen in Section 4.3, the effect of this resistive diffusion is always destabilizing, and the general result is that MHD instabilities can be excited [Antoni et al., 1986] and are indeed experimentally observed [Asakura et al., 1987; Watt and Nebel, 1983; Hirano et al., 1985; Antoni and Ortolani, 1987; Ueda et al., 1987; Cunnane et al., 1988; Tsui and Cunnane, 1988]. These resistive MHD instabilities may be the mechanism underlying the relaxation phenomenon because they allow the breaking and reconnection of magnetic field lines and lead to current redistribution [Caramana et al., 1983], thus restoring a relatively more stable profile in closer proximity to the fully relaxed minimum energy state. However, the fully relaxed $\beta = 0$ state cannot be accessed and a dynamic balance between relaxation and diffusion processes is achieved around a mean near-minimum energy state.

These concepts are heuristically illustrated in Figure 4-15. In this figure, level g is the "energy level" of an unconstrained thermal equilibrium (the "ground state" of the system). Level f is the "energy level" of the constrained, fully relaxed state. Level s represents a $\beta = 0$ MHD stable state with non-uniform μ, such as those discussed in Section 4.2. Finally, level p is the "energy level" of the partially relaxed state with finite β, representing the mean experimental state around which a driven system cyclically oscillates. Clearly, while levels g, f and s are not useful for plasma confinement, level p may very well be.

This picture of the dynamics of a relaxing profile is phenomenological and is based on independent analyses of mean field diffusion and linear MHD stability. The two results are conceptually combined in this dynamical cyclical model. However self-consistent 3D nonlinear MHD studies of the magnetic field time and space evolution have shown very clearly this behavior [Schnack et al., 1985; Kusano and Sato, 1987; Schnack and Ortolani, 1990], as discussed in Chapter 5. There is also a general agreement with observations and examples of the measured dynamics will be reviewed in the next section.

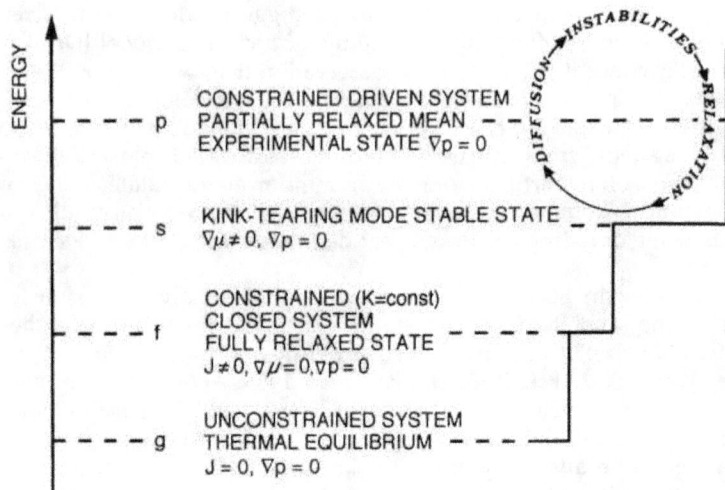

Figure 4-15. Schematic energy level diagram of the cyclical relaxation model.

In summary, the dynamics of the RFP magnetic field profiles occur about the near-minimum energy mean field profiles and can be described as being regulated by the competing processes of resistive diffusion and relaxation via MHD unstable perturbations.

4.5 Experimental Observations of Relaxation Phenomena in the RFP

We now turn our attention to a description of experimental results. Reversed-field pinch experiments have demonstrated the importance of relaxation phenomena. In particular, experimental measurements of fluctuations, large amplitude oscillations, and current density profile redistribution have shown that during both the RFP formation and sustainment phases, the relaxation process plays a determining role. Periodic oscillations of the RFP profile around a mean distribution have been observed [Watt and Nebel, 1983; Hirano et al., 1985; Asakura et al., 1987; Antoni and Ortolani, 1987; Ueda et al., 1987; Cunnane et al., 1988]. An example of the time evolution of the measured parallel current density profile is shown in Figure 4-16. Note the regular oscillations similar to those deduced from the phenomenological model described in Section 4.4.

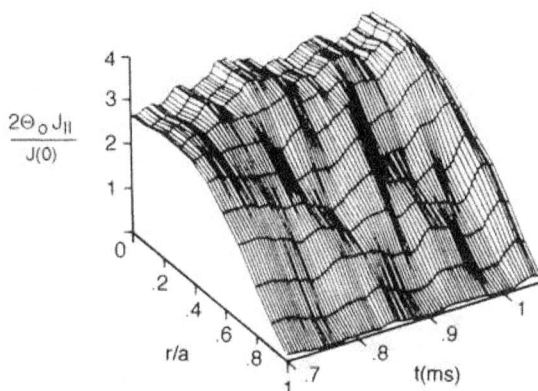

Figure 4-16. Measured current density profile dynamics.

In Figure 4-17 it is shown how these oscillations can be characterized both in terms of global parameters like Θ and α, and in terms of local quantities as Θ_0. Also the plasma thermodynamic quantities like density and temperature show regular oscillations as exemplified in Figure 4-18a,b for the soft x-ray radiation emission.

Although we have concentrated our discussion on relatively large amplitude, low frequency oscillations, a persistent broad frequency and mode spectrum of fluctuations is also observed. These fluctuations have an amplitude of the order of 1 per cent of the mean quantities, at frequencies around 10 kHz, poloidal periodicities $m = 0,1,2$ and toroidal periodicities covering the range $n = 3R/2a$ to $5R/a$. As an example, Figure 4-19 shows the power spectra, measured with edge coils, of B_z and B_θ at frequencies up to 200 kHz; it is seen that the power mostly occurs at low frequencies in the range 4 to 20 kHz and falls steadily as the frequency increases. Both field components behave in a similar way.

The analysis of the poloidal mode spectrum measured with coils at the edge of the plasma indicates that $m = 0$ and $m = 1$ are dominant, as shown in Figure 4-20 for fluctuations in the frequency range 4 to 20 kHz; data for B_θ are shown. A similar spectrum is obtained from B_z fluctuations. The dominant part is $m = 0$ and $m = 1$. Figure 4-21 shows the toroidal mode spectrum, which is broad, peaking at $n \sim -8$ (as expected for the aspect ratio of the experiment [*Hutchinson et al.*, 1984]; in general, as previously discussed, it is found that $n < n_0$ where n_0 represents the toroidal mode number associated with the on-axis $m = 1$ mode, $n_0 = -1/q_0 \sim -(3/2)(R/a)$. These $m = 1$ modes are

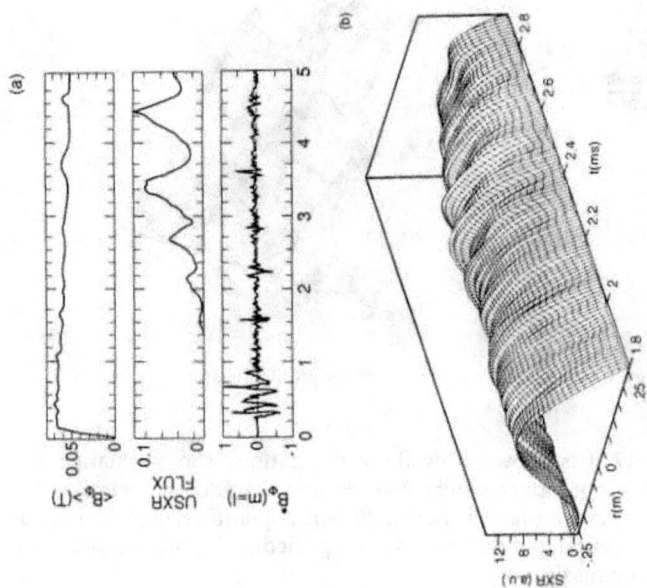

Figure 4-18. a. Waveforms of the average toroidal field, the ultrasoft x-ray (USXR) flux, and the field fluctuation amplitude for an $m = 1$ mode, showing sawtooth behavior [Watt and Nebel, 1983] b. Oscillations in the chord integrated SXR emission profile [Alper and Martin, 1989].

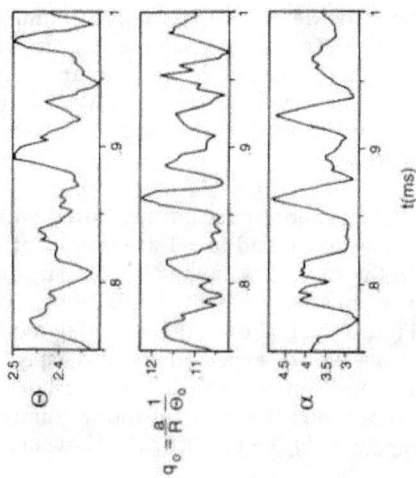

Figure 4-17. Measured oscillations of Θ, q_0 and α.

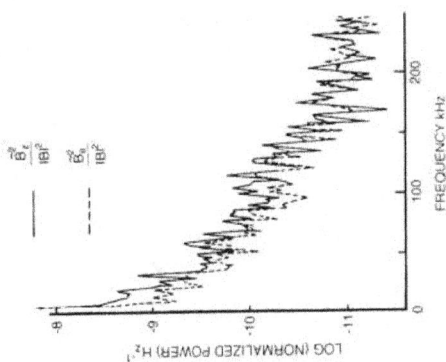

Figure 4-21. Toroidal mode spectra of magnetic field fluctuations in different frequency ranges obtained from experiment. The power is shown as a function of mode number, n [*Hutchinson et al.*, 1984].

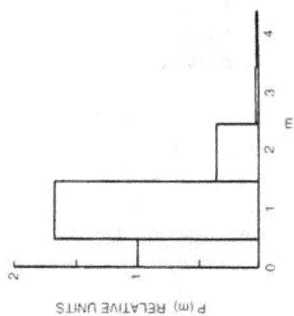

Figure 4-20. Poloidal mode spectrum of magnetic field fluctuations determined experimentally from $B\theta$. The power in the mode is shown as a function of mode number, m [*Hutchinson et al.*, 1984].

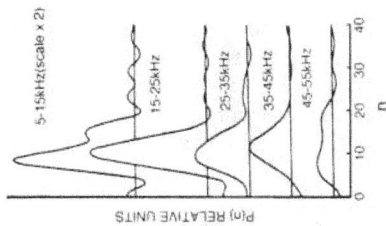

Figure 4-19. Frequency spectrum of magnetic field fluctuations observed on the toroidal and poloidal field components. [*Hutchinson et al.*, 1984]

found to be resonant in a region between the axis and the reversal surface, and correspond to the internal modes discussed in Section 4.2.

The short wavelength, localized activity that contains much of the power at high frequencies has been shown to have a short correlation length in the transverse direction. The radial cross correlation function is shown in Figure 4-22 for high and low frequency ranges, and it is seen that the former has a correlation length $\leq a/10$ (where a is the minor radius) compared with $\sim a$ for the latter. The longitudinal correlation length is long (at least $\sim R/2$).

Both large scale and localized modes occur throughout the discharge. In the current rise phase the amplitude of the large scale modes is much greater than during the sustainment phase, but the mode numbers and the frequency behavior change little. The universal occurrence of these modes is in agreement with the phenomenological model presented here, and suggests that they may play a fundamental role in the relaxation process.

Calculations for resistive tearing modes both from the linear theory discussed in this chapter and from the nonlinear analysis presented in the next chapter indicate that the most unstable modes have the same toroidal mode numbers as the observed $m = 1$ fluctuations. The radial variation of the amplitude of the perturbed field due to the instability has been determined by means of magnetic probes and an example of the results is shown in Figure 4-23. It is seen that the measured B_r component of the perturbed field due to the instability does not cross zero between the axis and the walls, which suggests a resistive MHD origin with $m = 1$. The calculated radial eigenfunctions are in reasonable agreement with the experimental observations.

As previously discussed the large scale $m = 1$ activity to be attributed to resistive modes implies a continuously evolving mode structure in a cyclic fashion. A mode grows, saturates and then decays by nonlinear mechanisms involving reconnection leading to profile modification, which leads to the excitation of modes with different n values and so on. The measured temporal evolution of the n spectra shown in Figure 4-24 lends support to this picture.

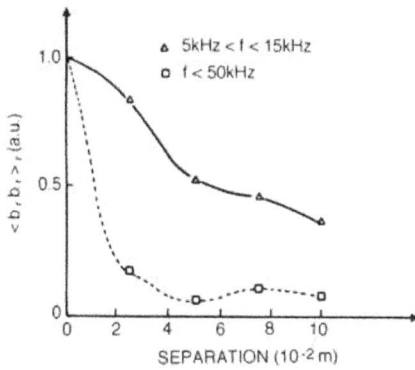

Figure 4-22. Radial cross correlation between the fluctuating radial magnetic field component, b_r, in the inner plasma regions and at other radial positions, for low and high frequencies.

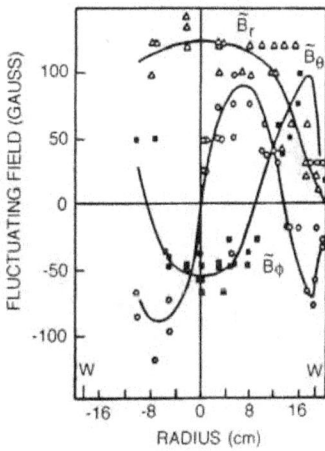

Figure 4-23. Radial variation of the fluctuating magnetic field components denoted by B_r, B_θ, B_ϕ, obtained from experiment. The wall is at W [LaHaye et al., 1984].

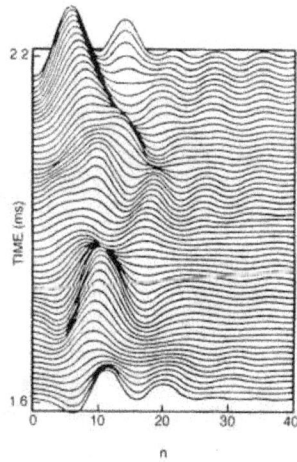

Figure 4-24. Time evolution of toroidal mode (n) spectrum during the current pulse. The peak of the spectrum varies in time as modes grow and decay followed by the growth of other modes of different n [Hutchinson et al., 1984].

CHAPTER 5

THE DYNAMICS OF PLASMA RELAXATION

Taylor's variational theory of plasma relaxation was reviewed in Chapter 3. This theory successfully predicts many of the global aspects of RFP (and other toroidal pinch) operation. However, it has several limitations. For example, it predicts completely relaxed states with constant normalized parallel current density, and uniform pressure. Also, it gives us no insight into the details of the dynamics that govern the relaxation process, other than that they be describable by the equations of resistive MHD. There is no simple formula or theory connecting the state variables, which describe the relaxed macroscopic equilibrium state, with the underlying dynamics (the fluctuations or turbulence). Instead, there is a body of evidence gained primarily from the numerical solution of the resistive MHD equations relating to plasma relaxation, especially in the RFP.

Phenomenological models and experimental results were reviewed in Chapter 4. A cyclic model for the RFP was developed. In this model the RFP exists in a constant state of tension between resistive diffusion and instability. This model was deduced from experimental observations and linear stability theory. It does not directly address the nonlinear dynamics responsible for restoring the relaxed profiles.

In this chapter, we will present the dynamical mechanism responsible for plasma relaxation as revealed by detailed nonlinear numerical simulations using the resistive MHD equations. The resulting picture has been successful in interpreting and understanding many global aspects of experimental RFP operation [Schnack, 1991]. We will see that, in agreement with the phenomenological models of Chapter 4, the RFP dynamo and plasma relaxation operate because of plasma instability. Within the discharge, there is a constant conflict between the forces of transport attempting to destroy the pinch, and the forces of instability attempting to restore the pinch to a more favored state. The RFP owes its existence to a stand-off between these competing processes.

In addition to describing the basic dynamical mechanism for sustainment, the MHD model can also *self-consistently* describe details of the spectra of magnetic fluctuations, the origin of anomalous plasma resistance, voltage and fluctuation increases in the presence of resistive walls and limiters, and sawtooth oscillations. These will be described in more detail in Chapters 6 and 7.

At the present time, the resistive MHD model is the *only* self-consistent picture of plasma relaxation. Because of the nonlinear, multidimensional nature of the RFP, much of the progress in the development of the dynamical model to be presented here has depended on the availability of advanced supercomputers, and the development of new techniques of numerical simulation. These issues were briefly discussed in Chapter 2. A more complete description of these techniques can be found in the literature [*Schnack et al.*, 1984, 1986, 1987].

We have noted that one of the characteristics of RFP plasmas is the maintenance of its magnetic field profiles for a time much longer than that corresponding to classical resistive diffusion. This has been called the *RFP dynamo*, in loose analogy with the magnetic field generation associated with the solar and terrestrial dynamos. We will begin this chapter with a brief description of the classical dynamo theory that has been developed to describe these naturally occurring phenomena. We will see that the RFP dynamo operates in a significantly different regime than the dynamos that occur in astrophysics and geophysics; the term "plasma relaxation" more appropriately describes the observations. Nonetheless, the term "RFP dynamo" has historical precedent, and is still used in the literature. In the first section of this chapter we shall attempt to clarify more precisely what occurs in the RFP, and point out how it differs from the astrophysical dynamo process.

5.1 Classical Dynamo Theory

The sustainment of the magnetic field in the RFP is reminiscent of phenomena observed in geophysical and astrophysical settings, where the magnetic fields of the earth and the sun also persist for anomalously long times, or appear in a quasi-periodic manner. It is natural, then, to look for generic mechanisms, consistent with Maxwell's equations, that can lead to the spontaneous generation of magnetic field in a conducting medium in such a way as to perpetuate magnetic fields in the presence of resistive (Ohmic) diffusion. Such naturally occurring processes are called *dynamos*, after the mechanical device that is used to generate electric current. Over the years, considerable theoretical effort has been expended in investigating the dynamo, almost all of it in an attempt to explain the astrophysical dynamo (see, for example, *Moffatt* [1978]; *Krause and Rädler* [1980]).

In this section we briefly discuss the elements of classical dynamo theory, as developed to explain astrophysical and geophysical observations. This will allow us to assess its relevance to the RFP dynamo, and to compare and contrast the physical processes occurring in the laboratory and astrophysical settings. We will see that the body of the theory is not directly applicable to the RFP because of the inherently different parameter ranges involved.   We review the theory primarily because of the original unsuccessful attempts to apply it to the RFP, to clarify some of the confusion this has caused, and because several of the concepts that arise will be useful later in describing toroidal pinch dynamics [Gimblett and Watkins, 1975].

Kinematic Dynamos

A theoretical description of the dynamo requires finding steady state (time independent) solutions of the resistive MHD Eqs. (2.27–2.30). We have seen that, in this model, Maxwell's equations in an appropriate form are coupled to the hydrodynamic equations through the inductive term in Ohm's law, which describes the effect of the fluid motion on the electric and magnetic fields, and the Lorentz force in the equation of motion, which describes the effect of these fields on the fluid motion. The combined system is nonlinear, and extremely difficult to analyze. It is therefore of interest to determine circumstances under which the fluid and field systems can be effectively decoupled.

Consider the forces acting on a fluid element, as given by the right-hand-side of Eq. (2.28). The Lorentz force acting on the fluid element, $J \times B$, can be ignored if it is small compared to the pressure force, $-\nabla p$. Assuming that the magnetic field and pressure vary on similar spatial scale lengths, this will occur if the condition $\beta \approx p/B^2 >> 1$ is satisfied, i.e., if the internal fluid energy density greatly exceeds the energy density of the magnetic field. The Lorentz force can also be ignored if it is small compared to the advective term $\rho v \cdot \nabla v$. This yields the condition $\rho v^2/2 >> B^2$; then the fluid kinetic energy density greatly exceeds the magnetic energy density.

If either of the above conditions is satisfied, the Lorentz force can be dropped from the momentum equation. In that case, the Maxwell and hydrodynamic equations become decoupled; the motion of the fluid proceeds independently from the development of the magnetic field. The theory then reduces to finding solutions to the field equations in the presence of a *prescribed* velocity field **v**. The only constraint placed on **v** is that it be *consistent* with the equation of motion.   Such flows are said to be *kinematically realizable*. (Theoretically, this condition is almost always supplemented by the auxiliary requirement $\nabla \cdot v = 0$.) The theory that results from this approximation is called *kinematic dynamo theory*.   The requirements of the preceding paragraph are always satisfied in the interior of the earth and stars [Moffatt, 1978].

The field Eqs. (2.17–2.20), can be combined into a single equation for the magnetic field **B**. In kinematic dynamo theory, we seek solutions of the resulting equation

$$\frac{\partial \mathbf{B}}{\partial t} = \nabla \times \left(\mathbf{v} \times \mathbf{B} - \frac{\eta}{S} \nabla \times \mathbf{B} \right),$$

(5.1)

with the functions $\mathbf{v}(\mathbf{r},t)$ and $\eta(\mathbf{r},t)$ assumed as given. Solutions of Eq. (5.1) qualify as dynamo solutions if the total magnetic energy

$$W = \frac{1}{2} \int B^2 dV$$

(5.2)

remains finite as $t \to \infty$.

It is important to recognize that Eq. (5.1) is *linear* in the magnetic field **B**. Kinematic dynamo theory is thus a linear theory, and is therefore amenable to analytic investigation. This possibly accounts for the theoretical concentration on this topic.

In many applications of interest, the spatial average of the magnetic fields is, at least on some length scale, symmetric with respect to an axis. This is illustrated in Figure 5-1a in cylindrical geometry (r, θ, z), where the fields are rotationally symmetric about the z-axis, and periodic in θ. An analogous case is shown in Figure 5-1b; the fields are now symmetric with respect to translations along the z-axis. These magnetic fields can be decomposed into poloidal and toroidal components, $\mathbf{B}_P$ and $\mathbf{B}_T$ [*Moffatt*, 1978]. For the case of Figure 5-1a, $\mathbf{B}_P = (B_r, 0, B_z)$ and $\mathbf{B}_T = (0, B_\theta, 0)$; for the case of Figure 5-1b, $\mathbf{B}_P = (B_r, B_\theta, 0)$ and $\mathbf{B}_T = (0, 0, B_z)$. These fields are topologically equivalent to the toroidal and cylindrical equilibria discussed in Section 1.3.

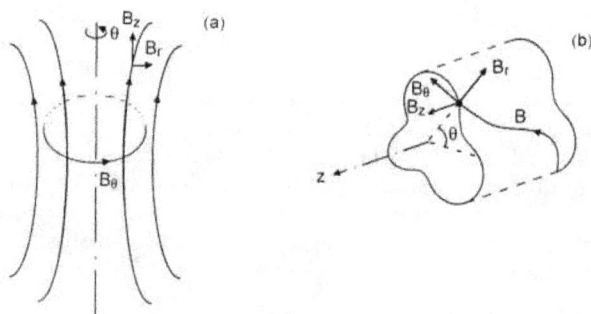

Figure 5-1. a. Rotationally symmetric magnetic field; b. Translationally symmetric magnetic field.

To obtain a steady configuration, one must identify motions that generate toroidal field from poloidal and *vice versa*; each field must provide the source for the other. It is easy to see that toroidal field can be generated out of poloidal field by imposing differential rotation about the axis of symmetry, or along the direction of symmetry [$v_\theta(r)$ for Figure 5-1a, $v_z(\theta)$ for Figure 5-1b]; the toroidal field is then generated simply by field line bending. The difficulties in the theory arise in identifying a method for generating poloidal field from toroidal field, as is required for a dynamo [*Moffatt, 1978*].

Cowling's Theorem

It is natural to seek solutions of Eq. (5.1) that correspond to the simplest possible velocity field **v**. Investigations of this sort have led to the unfortunate conclusion that dynamo solutions are impossible if **v** or **B** are too simple. These various results have become collectively known as *Cowling's Theorem*, after T. G. Cowling, who proved the first anti-dynamo theorem [*Cowling, 1934, 1957; Moffatt, 1978*]. This important result is that *magnetic fields that are symmetric about an axis cannot be maintained by symmetric motions.*

We shall prove a version of Cowling's Theorem in the special geometry of Figure 5-1b. The conclusions are, however, quite general, and applicable to a variety of magnetoplasma configurations.

For the geometry under consideration, the components of the poloidal field can be written in terms of a scalar flux function ψ as

$$B_r = -\frac{1}{r}\frac{\partial \psi}{\partial \theta}, \tag{5.3a}$$

$$B_\theta = \frac{\partial \psi}{\partial r}. \tag{5.3b}$$

With this *ansatz*, the z-component of Eq. (5.1) becomes

$$\frac{\partial \psi}{\partial t} + \mathbf{v}_P \cdot \nabla \psi = \lambda \nabla^2 \psi, \tag{5.4}$$

where $\lambda = \eta/S$, and $\mathbf{v}_p$ is the poloidal component of the velocity, which is assumed to satisfy $\nabla \cdot \mathbf{v}_p = 0$. We further assume that the normalized resistivity satisfies $\mathbf{v}_p \cdot \nabla \lambda = 0$, as would be the case if λ were merely convected with the fluid motion. Multiplying Eq. (5.4) by ψ/λ, integrating over all space V_∞, and integrating by parts, we find

$$\frac{d}{dt}\int_{V_\infty}\frac{\psi^2}{2\lambda}\,dV+\oint_{S_\infty}\left(\frac{\psi^2}{2\lambda}\mathbf{n}\cdot\mathbf{v}_P-\psi\mathbf{n}\cdot\nabla\psi\right)dS = -\int_{V_\infty}(\nabla\psi)^2dV\;,\qquad(5.5)$$

where S_∞ is the surface at infinity. The surface integral vanishes if $\psi \approx O(1/r)$ and $|\mathbf{v}_p|/\lambda \to 0$ as $r \to \infty$, or if $\mathbf{n}\cdot\mathbf{v}_p = \mathbf{n}\cdot\nabla\psi = 0$ on S_∞. These conditions are satisfied if the plasma is localized to a finite volume, or if S_∞ is a rigid, perfect conductor. The remaining volume integral on the right-hand-side of Eq. (5.5) is positive definite, so that ψ inevitably tends to zero as $t \to \infty$. The poloidal field thus cannot be maintained. The result also applies if we allow for an axial (toroidal) component of velocity (since $\mathbf{v}_T\cdot\nabla \equiv 0$); it depends only on the assumed translational symmetry of the magnetic field [Moffatt, 1978].

In addition to the inevitable decay of the symmetric poloidal field, it can be shown that, under the same conditions, the toroidal field must also decay [Moffatt, 1978]. Furthermore, it can be shown that dynamo action is impossible with purely toroidal motion, and with plane two-dimensional motion. Thus, the motions and magnetic fields associated with the dynamo are of necessity quite complicated, and therefore quite difficult to analyze. However, several specific examples of such dynamo solutions have been constructed, some of them quite ingenious [Moffatt, 1978]. Dynamo action is therefore at least possible, if not simple.

The consequences of Cowling's Theorem for the RFP can be understood heuristically [Gimblett, 1980]. Consider a symmetric cylindrical equilibrium with finite resistivity. If the toroidal field is reversed, there will be a cylindrical surface where $B_z = 0$ that lies within the plasma. On this surface a field line with $\mathbf{B} = B_\theta\hat{\mathbf{e}}_\theta$ closes upon itself in the poloidal (r,θ) plane. Along this field line the parallel electric field is strictly Ohmic. Thus the loop voltage around this surface is $\oint \mathbf{E}\cdot d\mathbf{l} = \oint \eta J_\theta r d\theta \neq 0$. The total toroidal flux contained within the reversal surface must therefore change with time, in contradiction to the assumption of equilibrium. Thus, a symmetric field-reversed state cannot be maintained; some sort of dynamo is required to sustain the discharge.

In the preceding, we have assumed that the velocity field $\mathbf{v}(r,t)$ is a known function of space and time. Solutions obtained under these conditions are known as *laminar dynamos*. We can also consider the case where $\mathbf{v}(r,t)$ contains a random component that is characterized only by its statistical properties. This random component will break the symmetry required for Cowling's Theorem to hold, and hence may admit the possibility of dynamo action. Such a component may result from small scale fluid turbulence, and the corresponding dynamo solutions are known as *turbulent dynamos*. These are briefly described in the following paragraphs.

The Turbulent Dynamo

We consider both the magnetic and velocity fields to consist of mean and random fluctuating parts. The fluctuating parts have the statistical property that their spatial and temporal averages vanish when considered on long enough space and time scales. It is also assumed that these space and time scales are short compared to the space and time scales of observational interest. We thus write

$$B(r,t) = B_0(r,t) + b(r,t) , \tag{5.6a}$$

$$B(r,t) = B_0(r,t) + b(r,t) , \tag{5.6b}$$

$$\langle b(r,t) \rangle = 0 , \tag{5.6c}$$

$$\langle u(r,t) \rangle = 0 , \tag{5.6d}$$

where $\langle .. \rangle$ represents an appropriate, and here unspecified, spatial and time average. With this assumption, Eq. (5.1) can also be separated into mean and fluctuating parts. We find that the mean fields evolve according to

$$\frac{\partial B_0}{\partial t} = \nabla \times (V_0 \times B_0) - \nabla \times E_f - \nabla \times \frac{\eta}{S} \nabla \times B_0 , \tag{5.7}$$

where

$$E_f = -\langle u \times b \rangle , \tag{5.8}$$

is a mean electric field arising from the interaction of the fluctuating velocity and magnetic field components.

It is important to note that E_f, being quadratic in fluctuating quantities, does not necessarily vanish. This quantity plays a central role in turbulent dynamo theory, and in the description of the RFP. Its presence on the right-hand-side of Eq. (5.7) allows for the possibility that E_f can balance the effects of resistive diffusion, even in cases where the laminar contribution satisfies the symmetry required for Cowling's Theorem.

The goal of turbulent dynamo theory is to identify the statistical properties of the fluctuating velocity field **u** that are consistent with dynamo action. A detailed description of this theory is beyond the scope of this book. Rather, here we concentrate on certain properties of Eq. (5.8) that will be relevant to our subsequent discussion of the RFP.

Note that, since Eq. (5.1) is linear in **B**, the fluctuating magnetic field **b** must be linearly related to the mean field B_0 [Moffatt, 1978]. It follows that E_f is also linear in B_0. We are thus motivated to express E_f as a series in B_0 and its derivatives. In Cartesian tensor notation, we write

$$E_{fi} = \alpha_{ij}B_{0j} + \beta_{ijk}\frac{\partial B_{0j}}{\partial x_k} + \gamma_{ijkl}\frac{\partial^2 B_{0j}}{\partial x_k \partial x_l} + \dots , \tag{5.9}$$

where α_{ij}, β_{ijk}, etc., are pseudo-tensors. The first term in the expansion is commonly called the α-effect, the second the β-effect, etc. Generally, these coefficients can be thought of as being properties of, or generated by, the turbulent velocity field.

Now consider the isotropic case $\alpha_{ij} = \alpha\delta_{ij}$, $\beta_{ijk} = \beta\varepsilon_{ijk}$, where α is a pseudo-scalar, β is a scalar, δ_{ij} is the Kronecker delta, and ε_{ijk} is the Levi-Civita symbol. Then the fluctuating electric field becomes $E_f = \alpha B_0 + \beta J_0$, and the mean component of Ohm's law is

$$E_0 = -V_0 \times B_0 + \alpha B_0 + (\eta/S + \beta)J_0 . \tag{5.10}$$

We can now see how the turbulent dynamo may work. Through the α-coefficient, the fluctuations generate a component of the mean electric field parallel to B_0 that can directly counterbalance the effects of resistive diffusion; since the fluctuations themselves are not symmetric, Cowling's Theorem does not apply, even though the mean field may be symmetric. This is called the α-effect. Through the β-coefficient, the effective resistivity may be directly changed. This is called the β-effect. We shall see that the RFP dynamo can be described generally in terms of the α-effect.

Relevance to the RFP Dynamo

We remark on some important differences between the RFP dynamo and the classical astrophysical or terrestrial dynamo that have been the source of some confusion. As we have seen, classical dynamo theory is concerned with the generation of magnetic field in the presence of a kinematically realizable velocity field, which is taken to be given *a priori* and fixed for all time. (This last statement may apply to the velocity field itself, or to its statistical properties.) It is assumed that the magnetic field has little or no effect on the velocity field, so that the equation of motion need not be solved. This leads to a linear theory for the magnetic field. Such models are valid in the limit where either the kinetic energy or the thermodynamic energy greatly exceeds the magnetic energy, as is the case for geophysical and astrophysical dynamos.

The RFP exists in the opposite limit: the magnetic field energy greatly exceeds the kinetic energy, and $\beta = 8\pi p / B^2 < 1$. The equation of motion must then be solved simultaneously with the induction equation and Ohm's law. The RFP dynamo is thus inherently *nonlinear*. Furthermore, the classical dynamo concept requires the cyclic conversion of poloidal flux into toroidal flux, and *vice versa*. The former can be achieved trivially by differential rotation. The difficulties in the theory arise in the regeneration of poloidal flux from toroidal flux. In the RFP the poloidal flux is continually supplied by the external circuit; there is no need for a reconversion mechanism. If the loop voltage is set to zero, the RFP dynamo eventually ceases to operate and the discharge terminates by resistive diffusion. The term "dynamo" used to describe the RFP is therefore valid in the sense that condition given by Eq. (5.2) is satisfied and that the reversed field is sustained and regenerated. It originated in the early days of laboratory pinch research when the process was not well understood. It is not surprising that attempts to apply classical dynamo theory directly to the RFP have met with little or no success.

Nonetheless, certain concepts developed in turbulent dynamo theory are useful in describing the dynamics of the RFP. While the RFP does not behave as a kinematic dynamo, it does exhibit anomalously long field lifetimes. Thus, the role of fluctuations in generating a mean electric field parallel to B_0, that can counteract the effects of resistive diffusion, is important. The fluctuations that are important in the RFP are not small scale turbulence, but are rather a superposition of long wavelength laminar motions. Thus, the detailed results of turbulent dynamo theory relating the expansion coefficients of Eq. (5.9) to the turbulence are generally not relevant. However, the effect of the fluctuations on the mean field can still be expressed in a form given by Eq. (5.10). We shall continue to make reference to the α-effect throughout the remainder of this book.

Finally, we comment on the inherent nonlinearity of the RFP dynamo. A variational description of global sustainment in toroidal pinches has been developed [*Taylor*, 1974, 1986], and has been described in Chapter 3. However, nonlinearity has made the *details* of the RFP dynamo virtually inaccessible by standard analytic techniques. Progress in understanding this process has required the application of advanced numerical algorithms, and advances in high speed supercomputer technology. The results of these calculations are described in the following sections.

5.2 The Basic Relaxation Mechanism

The basic mechanism of field reversal and sustainment in the RFP is the nonlinear evolution of $m = 1$ kink instabilities in the presence of an externally applied toroidal voltage. The helical displacement produced by such modes results in a tilting of poloidal magnetic field components into the axial (toroidal) direction, thus converting poloidal field into toroidal field in a manner analogous to differential rotation. If the axial current is large enough, the instability grows with the proper pitch (handedness) to an amplitude that is sufficient to reverse the average axial field at the edge of the plasma. After saturation, this average reversal decays, but before reversal is lost another kink mode is destabilized, thus reinforcing the negative axial field at the edge. The free energy for these modes is supplied by the Poynting flux from the applied voltage, and by the deviation of the mean field profiles from the Bessel function model.

The Original Work of Sykes and Wesson

The physical process described above was first recognized by Sykes and Wesson [Sykes and Wesson, 1977] as a result of an early application of large scale numerical simulation to a pinch configuration. They used an explicit numerical algorithm to study the evolution of the resistive MHD equations in a periodic, square cross-section cylinder in response to an externally applied axial electric field. The axial flux was held constant throughout the calculation. The length of the box was consistent with a toroidal aspect ratio $R/a = 1/2$. Finite differences were used for all spatial discretization. Computational resources of the time limited the spatial resolution to a $14 \times 14 \times 13$ grid. Nonetheless, these calculations exhibited strong dynamo action, indicating the robust nature of the relaxation process in the RFP.

The results of this simulation are illustrated in Figures 5-2(a-c) [Sykes and Wesson, 1977]. For these results, time is measured in units of the Alfvén transit time $\tau_A = a/V_{A0}$, where $V_{A0}^2 = B_0^2/4\pi\rho$. In Figure 5-2a the axial voltage and current are plotted as functions of time. The current is raised rapidly from $t = 0$ simulating a rapidly formed discharge; the voltage shown is what is required to give the current waveform. Figure 5-2b shows the time variation of the average axial magnetic field at the edge of the plasma, measured in units of the initial field on axis. It can be seen that for times up to $t = 5t_A$, $\langle B_{za} \rangle$ is considerably reduced due to plasma compression and axial flux conservation. Up to this time the plasma column has remained almost axisymmetric. However, at $t = 4.8\tau_A$, a single helical instability with axial wavelength equal to the box length is growing. The pitch (handedness) of this mode corresponds to an axial mode number $n = -1$.

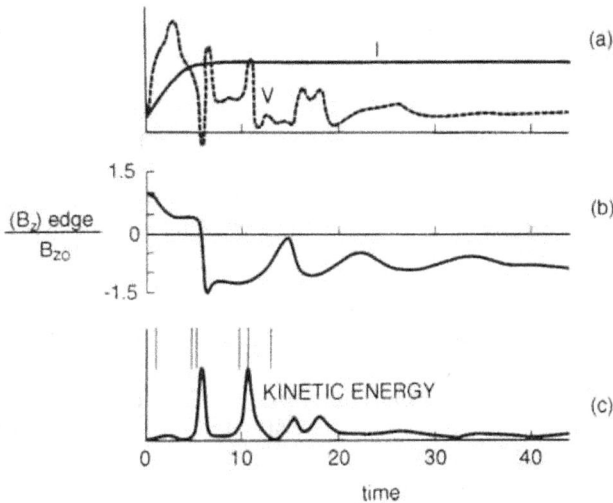

Figure 5-2. a. Axial voltage (V) and total current (I) vs. time; **b.** Normalized axial magnetic field at the plasma edge vs. time; **c.** Kinetic energy vs. time. Time is measured in units of the Alfvén transit time. The sustained field reversal beyond $t = 15$ is evidence of the dynamo [*Sykes and Wesson*, 1977].

In Figure 5-2c we display the kinetic energy as a function of time. Just after $t = 5\tau_A$, the kinetic energy grows rapidly, and the axial magnetic field become strongly reversed. This is due to the nonlinear growth of the helical instability causing a helical distortion of the current channel. At $t = 9\tau_A$, the helically deformed channel becomes again unstable to another helical mode with half the wavelength of the initial instability, i.e., an $n = -2$ mode. This is clearly seen in the growth of kinetic energy at $t = 9.8\tau_A$. The result of the nonlinear interaction of the $n = -1$ and $n = -2$ modes is to restore the plasma to an almost axisymmetric state, but with a reversed axial field. This state is presumably more stable than the original non-reversed state, and is the preferred relaxed state of the discharge. Because of its stability, the plasma is no longer dynamical, but begins to resistively decay, with field reversal being almost lost at $t = 14\tau_A$. However, this loss is prevented by the occurrence of further instability; field reversal is apparently maintained indefinitely by a series of oscillations of the type just described. As originally pointed out by

Sykes and Wesson, the key question remains not whether field reversal is maintained, but rather the level of energy transport that may result from the unstable fluctuations required to maintain field reversal. These issues are further addressed in Chapter 7.

Spontaneous and Driven Reconnection in the RFP

Experimentally, it is observed that there is a limit to the amount of current density that can be sustained in the core of the RFP. This limit is such that $q(0) \gtrsim 2a/3R$, and quasiperiodic oscillations in $q(0)$ near this value have been observed. This was described in Chapter 4, where it was suggested that this limit is imposed by instability of $m = 1$ modes. In this section we consider the q profile of an RFP, and its evolution due to resistive diffusion [Caramana et al., 1983].

As indicated in Section 2.3 and described in Chapter 4, the RFP q profile is a monotonically decreasing function of radius, with $q(0) \approx a/R$ and $q(a) < 0$. We have also seen that the detailed shape of the q profile can affect the stability of the discharge to ideal and resistive MHD modes. We can now heuristically envision how the value of $q(0)$ might evolve in response to resistive diffusion and instability.

Let us consider the forces that are at work to alter the preferred, reversed field profile. These forces were briefly considered in Section 4.4. It is likely that RFP plasmas have a cold edge, and therefore the resistivity is larger in the outer regions of the discharge than in the center. Under these circumstances a global thermal instability driven by nonuniform Ohmic heating can occur if the resistivity varies inversely with some power of the temperature. Because the central plasma is a better electrical conductor than the edge plasma, the current preferentially flows near the magnetic axis. The resulting Ohmic dissipation preferentially heats this region, thereby raising the central temperature, lowering the central resistivity relative to that at the edge, and further channeling the current to the axis. Thus the central current density will increase in time. Simultaneously, resistive diffusion in the cold edge region begins to destroy the reversed axial field. Left to its own devices, the original magnetic field profiles would become greatly modified, and would cease to resemble an RFP.

The thermal instability will cause $q(0) \approx B_z(0)/J_z(0)$ to fall. As $q(0)$ falls, the q profile in the central region of the discharge is flattened, thus decreasing the stabilizing effect of shear (q'/q). As a result, a global MHD kink mode with poloidal mode number $m = 1$ and toroidal (axial) mode number n, that is either a resistive instability resonant near the axis $[0 < q_{res} (= -m/n) < q(0)]$, or an ideal instability closely nonresonant from above $[q(0) < q_{res}]$, is destabilized. This mode grows faster than the thermal instability process driven by transport. If the mode is resonant, its nonlinear evolution will result in magnetic reconnection of the type discussed in Section 2.2. One result of this global reconnection process is to remove the original resonance from the plasma [*Kadomtsev*, 1975], i.e., $q(0)$ is further lowered until $q(0) < -m/n$, i.e., the mode becomes nonresonant from above. However, this removes more shear from the profile, leaving the plasma in an even more unfavorable state. The original (m,n) mode responds by continuing its robust nonlinear evolution through a process that involves a second driven reconnection (i.e., reconnection as a result of imposed plasma flows, rather than linear instability) that reintroduces the original resonance into the plasma and raises $q(0)$ to a level greater than its original value. Raising $q(0)$ is equivalent to flattening the parallel current profile outward, thus mitigating the effects of the thermal instability, driving poloidal current in the plasma edge, and restoring a stable RFP profile. The process is then free to repeat in a cyclic manner.

If the mode originally destabilized by transport processes is the ideal kink mode that is closely nonresonant above, only the second nonlinearly driven reconnection occurs, and the stable RFP profile is again restored.

The dynamics described above are best illustrated by direct numerical solution of the resistive MHD equations [*Schnack et al.*, 1985]. In Figure 5-3 we plot $q(0)$ versus time during the nonlinear, single helicity (two-dimensional) evolution of an $m = 1$ resistive kink mode originally resonant near the axis at $q_{res} = 0.1$. Note that, due to magnetic reconnection, $q(0)$ is decreased until the resonant value appears at the axis. This is the first reconnection. The subsequent rapid increase in $q(0)$ to a value larger than its original value is a result of the second, nonlinearly driven reconnection. The final profile is substantially more stable than it was at $t = 0$.

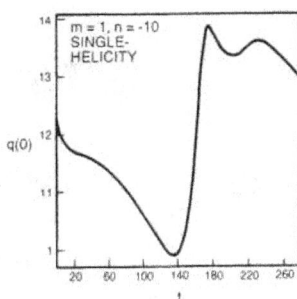

Figure 5-3. $q(0)$, the safety factor at the magnetic axis, vs. time during the growth and saturation of an $m = 1$, $n = -10$ kink mode in single helicity. Time is measured in units of the Alfvén transit time.

The evolution of the appropriate helical flux surfaces during this process [*Caramana et al.*, 1983] is illustrated in Figures 5-4(a-f). The first reconnection (Figures 5-4(a-c)) removes the resonance, while the second (Figures 5-4(d-f)) restores it. When the unstable mode is originally nonresonant from above, only the second rapid increase in $q(0)$ occurs. The flux surfaces in this case are shown in Figure 5-5. The field reversal parameter F for a similar case [*Schnack et al.*, 1985] is shown as a function of time in Figure 5-6. Note that, in this case, field reversal is attained and maintained for a large fraction of the global resistive diffusion time, illustrating the robust operation of the RFP dynamo.

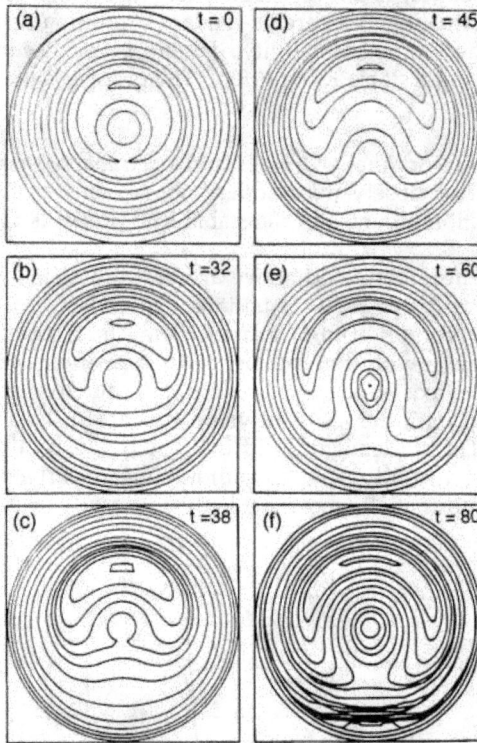

Figure 5-4. Nonlinear evolution of helical flux surfaces during the single helicity growth and saturation of an internally resonant $m = 1$ kink mode. The first reconnection (a-c) removes the original resonance, as evidenced by the disappearance of the magnetic island. The second reconnection (d-f) restores both the resonance and the island, but leaves the plasma in a stable state.

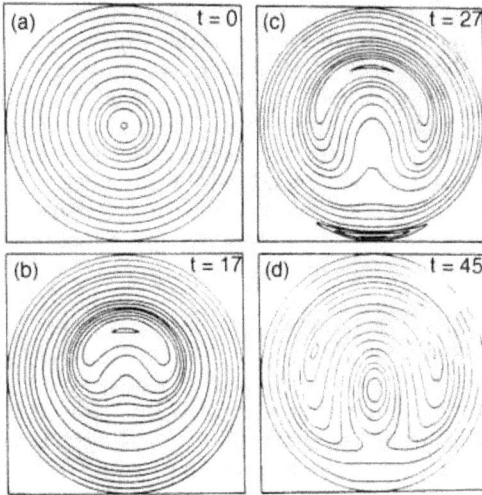

Figure 5-5. Nonlinear evolution of helical flux surfaces during the single helicity growth and saturation of an $m = 1$ kink mode that is nonresonant from above. Driven reconnection (c,d) introduces the resonance and its associated magnetic island.

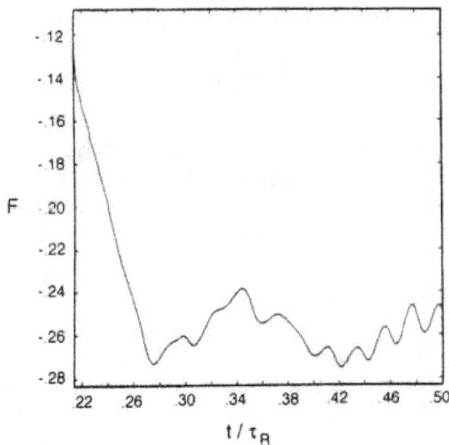

Figure 5-6. Field reversal parameter $F = B_z(a)/\langle B_z \rangle$ vs. time from a three-dimensional numerical simulation. Time is measured in units of the resistive diffusion time.

To summarize, the basic RFP dynamo mechanism proceeds as follows. Global $m = 1$ kink modes are driven unstable by current peaking due to transport processes that *lower* $q(0)$. If the primary unstable mode is resonant, its nonlinear evolution proceeds in two steps: a *first reconnection* that removes the original resonance and further lowers $q(0)$; and, a *second reconnection* that restores the original resonance, *raises* $q(0)$, and restores a stable profile. The second reconnection is a driven, nonlinear process. If the primary unstable mode is nonresonant from above, only the second reconnection occurs. These are the dynamics that underlie the cyclical process discussed in Sections 4.4 and 4.5.

Fluctuations and Ohm's Law

The dynamo mechanism first identified by Sykes and Wesson [1977], and described in the preceding paragraphs [*Caramana et al.*, 1983] clearly works to oppose the action of thermal instability and resistive diffusion. A different view of the basic dynamo mechanism is obtained by considering Ohm's law, Eq. (2.20). As discussed in Section 5.1, in a fluctuating plasma the magnetic and electric fields in Eq. (2.20) contain both mean (average) and oscillating parts. Assume that the velocity field has only a fluctuating component. Then taking the spatial average of Ohm's law, we find that the mean ($m = 0$, $n = 0$, in a cylindrical cross-section discharge) electric field is given by

$$E_0 = E_f + \frac{\eta}{S} J_0 , \qquad (5.11)$$

where E_f is the mean part of the electric field produced by the nonlinear interaction of the fluctuating velocity and magnetic fields,

$$E_f = -\langle v \times b \rangle . \qquad (5.12)$$

Here, we have set the mean flow $V_0 = 0$.

Now consider the component of Eq. (5.11) parallel to the mean field B_0. (This is formally analogous to the α-effect dynamo discussed in Section 5.1.) In a temporal steady state, this is just the parallel component of the spatially uniform applied toroidal electric field. In Figure 5-7 we plot the fluctuation-induced and resistive parts of the parallel component of Eq. (5.11) as a function of radius during such a steady state when the RFP dynamo is fully active. These results were obtained from a fully three-dimensional, well resolved numerical solution of the force-free model given by Eqs. (2.58) with a close fitting conducting shell [*Ho et al.*, 1989]. Positive values of $E_{f\parallel}$ serve to *suppress* parallel current, while negative values serve to *drive* parallel

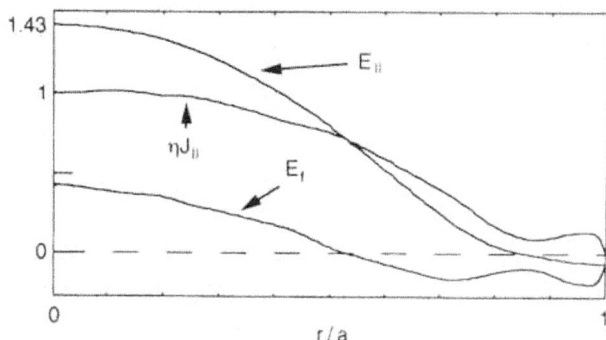

Figure 5-7. Average parallel electric fields vs. radius during the sustainment phase of a three-dimensional numerical simulation of an RFP discharge with a perfectly conducting outer boundary. Ohmic, fluctuating (dynamo), and total fields are shown. Fields are normalized to the Ohmic electric field on axis. In the absence of fluctuations, the field would be entirely Ohmic; the fluctuations result in a total electric field that is 43% anomalous on axis.

current. (Recall that, in the absence of fluctuations, only the resistive part of Eq. (5.11) can contribute to E_0.) We see that the effect of the parallel fluctuation-induced electric field is to suppress axial current on the axis, and to drive poloidal current at the edge. But this is just what is required to counteract both the current peaking driven by the thermal instability and the resistive diffusion of the negative field at the plasma edge. In steady state, a perfect balance is reached and the dynamo sustains the discharge as long as the external voltage is supplied (i.e., E_0 is maintained constant.)

It has also been demonstrated [*Schnack et al.*, 1985; *Ho et al.*, 1989; *Nebel et al.*, 1989; *Caramana*, 1989] that only the $m = 1$ contribution to E_f produces the dynamo; $m = 0$ and $m > 1$ give either insignificant or anti-dynamo contributions. Furthermore, $m = 1$ modes alone are sufficient to produce dynamo action [*Caramana*, 1989].

Evidence of Taylor Relaxation

Taylor's relaxation theory requires that the magnetic energy W be minimized while the magnetic helicity K remains invariant, i.e., the ratio W/K should seek a minimum. In Figures 5-8(a,b) we plot the ratio W/K, and F as functions of time during the numerical simulation of the attainment and sustainment of field reversal in an RFP discharge with a perfectly conducting outer boundary. It is clear that W/K is minimized during the relaxation

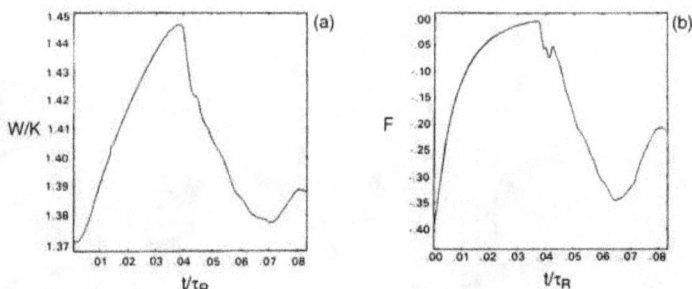

Figure 5-8. Time evolution of **a.** the Taylor ratio W/K, and **b.** the field reversal parameter F, from a three dimensional numerical simulation of RFP sustainment. Time is measured in units of the resistive diffusion time. W is the magnetic energy and K is the total helicity. The decrease in W/K is evidence of plasma relaxation associated with field reversal.

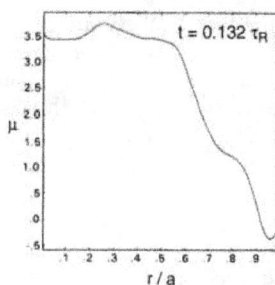

Figure 5-9. The normalized parallel current density μ vs. radius during the sustainment phase of a three-dimensional numerical simulation of an RFP discharge. Note that μ is constant for $r/a < 0.6$, indicating relaxation in the core of the discharge. The "knee" at $r/a = 0.8$ is a quasilinear modification due to the presence of $m = 0$ modes.

process, thus verifying a posteriori the original insight of Taylor [*Taylor*, 1974]. In Figure 5-9 we plot the radial profile of the normalized parallel current $\mu(r) = \mathbf{J} \cdot \mathbf{B} / B^2$ that results from such a simulation. Recall that the Taylor theory predicts $\mu = constant$ (see Section 3.4). We see that the dynamo process has produced such a state over the inner half of the discharge, with a value consistent with approximately Bessel function fields. Complete relaxation in the sense of Taylor is thus apparently not necessary to yield a sufficiently stable state. Note that the profile in Figure 5-9 is similar to the experimental results phenomenologically described in Chapter 4.

The Helical Ohmic State

We have seen that the RFP dynamo arises from the nonlinear evolution of helical kink instabilities coupled with slow profile modification by transport processes. It is interesting that helically symmetric resistive steady states with plasma flow and externally applied voltage can be computed directly as an equilibrium problem [*Dobrott et al.*, 1985]. These helical ohmic states [*Wesson*, 1979] represent the essential feature of the RFP dynamo.

In the zero-beta (pressureless) limit, the relevant equations of the model are

$$\nabla \cdot \mathbf{B} = 0 , \qquad (5.13a)$$

$$\nabla \times \mathbf{B} = \mathbf{J} , \qquad (5.13b)$$

$$-\nabla \Phi + \mathbf{v} \times \mathbf{B} = \eta \mathbf{J} , \qquad (5.13c)$$

$$\mathbf{J} \times \mathbf{B} = 0 , \qquad (5.13d)$$

where the electric field has been written as $\mathbf{E} = -\nabla \Phi$, with $\Phi = E_0 z \, \hat{z} + \phi$. Here E_0 is the applied toroidal electric field and ϕ is a periodic potential function. It is assumed that all quantities depend on the cylindrical coordinates (r, θ, z) only through the radius r and the helical variable $u = l\theta + kz$, where l/k is a fixed quantity that specifies the pitch of the helical coordinate system. Then by Eq. (5.13a), the magnetic field can be written as

$$\mathbf{B} = \hat{\mathbf{w}} I - \hat{\mathbf{w}} \times \nabla \psi , \qquad (5.14)$$

where $I = I(r, u)$, and $\psi = \psi(r, u)$ is the helical flux. The unit vector $\hat{\mathbf{w}}$ is related to the helical unit vector

$$\hat{\mathbf{u}} = \frac{l\hat{\theta} + kr\hat{z}}{\sqrt{l^2 + k^2 r^2}} , \qquad (5.15)$$

by

$$\hat{\mathbf{w}} = \hat{\mathbf{r}} \times \hat{\mathbf{u}} . \qquad (5.16)$$

With this *ansatz*, along with the assumptions of small flow and resistivity, Eqs. (5.13) can be reduced to a 1-1/2 dimensional system analogous to the familiar Grad-Shafranov equation of MHD equilibrium [*Bateman*, 1978; *Freidberg*, 1987]. The resulting system is

$$\nabla \cdot (h^2 \nabla \psi) + 2klh^4 I + h^2 II' = 0 , \tag{5.17a}$$

$$l \left\langle h^2 \right\rangle IE_0 - k \left\langle rh^2 \frac{\partial \psi}{\partial r} \right\rangle = \left\langle \eta h^2 \right\rangle I^2 I' + I' \left\langle \eta h^2 |\nabla \psi|^2 \right\rangle , \tag{5.17b}$$

where $I = I(\psi)$, $h^2 = 1/(l^2 + k^2 r^2)$, $\langle ... \rangle$ denotes a flux surface average, and $(..)'$ indicates differentiation with respect to ψ. Equations (5.17(a,b)) can be solved numerically by an iterative procedure once the resistivity profile $\eta(r)$ and the applied electric field E_0 are specified [*Dobrott et al.*, 1985].

Helical ohmic states can also be computed as the time asymptotic steady states of time-dependent simulations [*Aydemir and Barnes*, 1984; *Dobrott et al.*, 1985; *Montgomery*, 1989; *Finn et al.*, 1992]. These calculations used the force-free model described in Chapter 2, and were limited to the symmetry containing a single helicity defined by the mode number (m,n) of the dominant dynamo mode. It is of interest to compare these states with those obtained from the direct solution of Eq. (5.17(a,b)). When the same aspect ratio, resistivity profile, and applied electric field are used [*Dobrott et al.*, 1985], a state with $\Theta = 1.71$, $F = -0.23$ is obtained with the time-dependent code, and $\Theta = 1.71$, $F = -0.26$ is obtained from Eq. (5.17(a,b)). In Figure 5-10 we plot the helical flux surfaces $\psi = constant$ obtained from Eq. (5.17(a,b)). The flux surfaces obtained from the time-dependent calculation are indistinguishable from these.

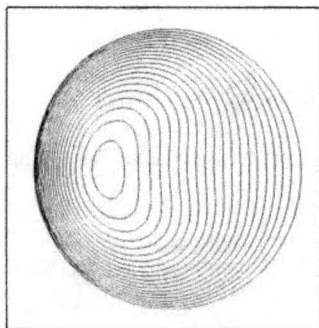

Figure 5-10. Helical flux surfaces in a helical Ohmic state obtained from numerical solution of Eqs. (5-17(a,b)) [*Dobrott et al.*, 1985].

We comment that these helical states may be themselves unstable to modes of a different symmetry (pitch). This is similar to the second instability observed by Sykes and Wesson.

Helical ohmic states similar to those described here have recently been shown to be states of minimum dissipation [*Montgomery and Phillips*, 1988]. It has been speculated [*Montgomery*, 1989] that these solutions of the resistive MHD equations may play a fundamental role in describing the possible states of real (non-ideal) plasmas, similar to the role played by Poisseuille and Couette flows in viscous hydrodynamics [*Landau and Lifshitz*, 1959].

5.3 Effects of Nonlinear Mode Coupling

We have seen that the basic mechanism of the RFP dynamo arises because of the nonlinear evolution of unstable $m = 1$ MHD kink modes [*Sykes and Wesson*, 1977; *Caramana et al.*, 1983; *Schnack et al.*, 1985]. Quasilinear calculations have shown that a single such mode is sufficient to sustain the discharge [*Caramana*, 1989]. Furthermore, the sustained dynamo state can be cast as an Ohmic equilibrium with flow that exists in a perfect steady state for all time. Unfortunately, such steady states are not observed experimentally. Instead, as described in Chapter 4, the RFP typically exhibits a variety of persistent dynamical behavior ranging from relatively small scale magnetic fluctuations to coherent sawtooth oscillations. Furthermore, this behavior is observed to vary qualitatively with different operating parameters and boundary conditions.

In this section we present the results of large scale numerical simulations of RFP dynamics at experimentally relevant aspect ratio. These calculations include the interaction of many unstable normal modes. In this way a more realistic, dynamical picture of the RFP emerges that still retains the aspects of the basic RFP dynamo mechanism.

MHD Fluctuations

The RFP dynamo mechanism presented in Section 5.2 is basically independent of the pinch aspect ratio R/a, and persists robustly at $R/a = 1$. Recall that for an RFP $q(0) \approx a/R$, and that for $m = 1$ kink modes the resonance occurs at $q_{res} \approx 1/|n|$, where n is the axial mode number of the instability. Thus the locations of the resonances of successive modes $|n| = 1, 2, ...$, become more closely spaced as $|n|$ gets large, becoming dense at the field reversal surface $q = 0$ as $|n| \to \infty$. These issues were discussed in more detail in Section 2.3. For unit aspect ratio, we expect a low $|n|$ mode to be the most unstable. The mode with the next highest mode number is resonant far from the axis ($r = 0$), and is likely to be stable. Thus the preferred helical distortion is determined by the single unstable kink mode; a helical ohmic state can exist with this helical pitch.

Now consider a more realistic aspect ratio, $R/a \approx 3\text{-}5$, say. Then there will be more unstable resonant modes existing near $r = 0$; their resonances will be more closely spaced. Each of these unstable modes is capable of producing a helical ohmic state. But each of these states may itself be unstable to neighboring modes of a slightly different pitch. There are thus many possible helical deformations, none of which is preferred. As a result of this mode competition, the plasma oscillates between these states, producing the MHD fluctuations that are experimentally observed. This is illustrated in Figure 5-11, where we plot the energy in several modes as a function of time for a case [*Schnack and Ortolani*, 1990] with aspect ratio $R/a = 4$. Several $m = 1$ modes are initially unstable; in the saturated state no single mode is observed to dominate. In Figure 5-12 we plot F versus time for a similar case; note the oscillations after field reversal is attained, indicating ongoing fluctuations. Similar oscillations occur in $q(0)$, as shown in Figure 5-13a. Nonlinear mode coupling to $m = 0$ may enhance this process. This will be described in the following paragraphs. These oscillations closely resemble those observed experimentally [*Antoni and Ortolani*, 1987], as described in Chapter 4. In Figure 5-13b we show the corresponding evolution of W/K, indicating the ongoing relaxation process.

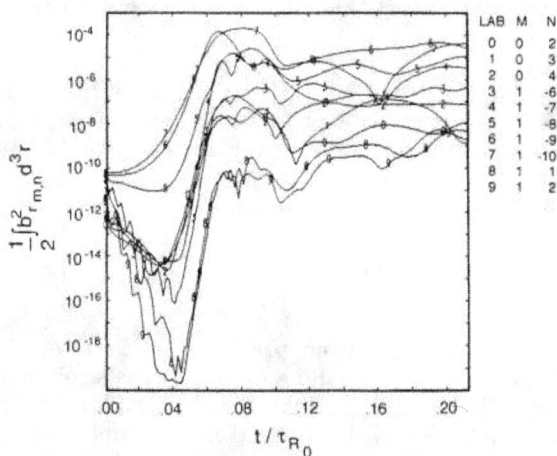

Figure 5-11. Energy in the radial magnetic field of selected modes vs. time from a three dimensional simulation of RFP reversal and sustainment. The aspect ration is $R/a = 4$. Note that many modes are unstable. No single mode dominates during the sustainment phase, $t/\tau_R > 0.14$.

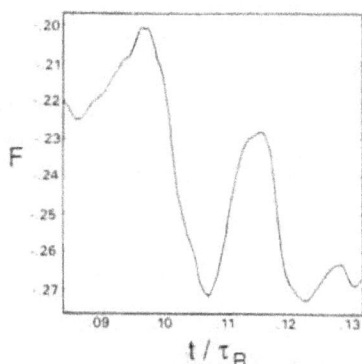

Figure 5-12. Field reversal parameter F vs. time from a three dimensional simulation of the sustainment phase of a large aspect ratio RFP. The oscillations are associated with relaxation events.

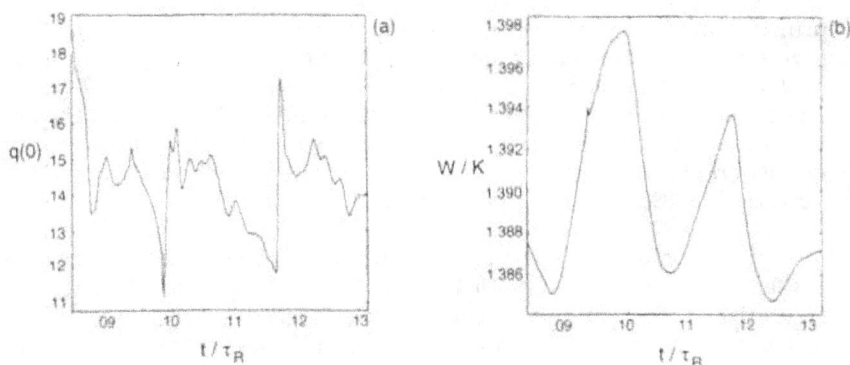

Figure 5-13. a. $q(0)$, the safety factor on axis, and **b.** the Taylor ratio W/K, vs. time during the sustainment phase shown in Figure 5-12. Generation of reversed field is associated with rapid increases in $q(0)$ (as in Figure 5-3) and decreases in W/K, indicating relaxation. The oscillations in $q(0)$ are related to oscillations in the on-axis current density. (Compare with experimental measurements shown in Figure 4-16.)

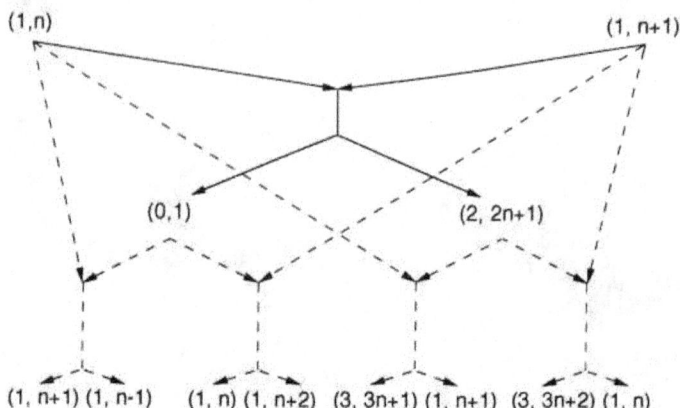

Figure 5-14. Mode coupling diagram for nonlinear dynamics in the RFP.

Nonlinear Mode Coupling

Since the RFP is fundamentally a three-dimensional nonlinear dynamical system, it is expected that mode coupling will play a role in determining the character of the observed fluctuations. At realistic aspect ratio, several $m = 1$ modes with $|n| > 1$ will likely interact nonlinearly. The primary mode couplings to be expected from such an interaction [*Caramana et al.*, 1983] are a quasilinear coupling to $m = 0$, $n = 0$ (the mean fields), and a nonlinear coupling to both $m = 0, |n| \approx 1$, and $m = 2, |n| \gg 1$. These mode couplings are illustrated schematically in Figure 5-14. The details of these couplings have been the subject of several studies [*Schnack et al.*, 1985; *Strauss*, 1985; *Kusano and Sato*, 1987; *Holmes et al.*, 1988; *Nebel et al.*, 1989; *Caramana*, 1989; *Ho and Craddock*, 1991].

Recently, a picture of a plausible mode coupling mechanism in the RFP has emerged [*Ho and Craddock*, 1991]. This is shown schematically in Figure 5-15. The external circuit continually feeds energy into the mean ($m = 0$, $n = 0$) profiles; this is the energy that drives the dynamo. Resistive diffusion destabilizes $m = 1, |n| > 1$ modes that are resonant near the axis. These are the essential dynamo modes discussed in Section 5.2. They serve to quasilinearly relax the mean profiles. The nonlinear interaction of these modes drives both $m = 0, |n| \approx 1$, and $m = 2, |n| \gg 1$ modes. All of these modes are linearly stable. The $m = 2$ modes simply serve as an energy sink, eventually giving up their energy by cascading to shorter wavelength and resistive dissipation [*Holmes et al.*, 1988]. However, in addition to providing an energy sink, the $m = 0$ modes play an important role in determining the

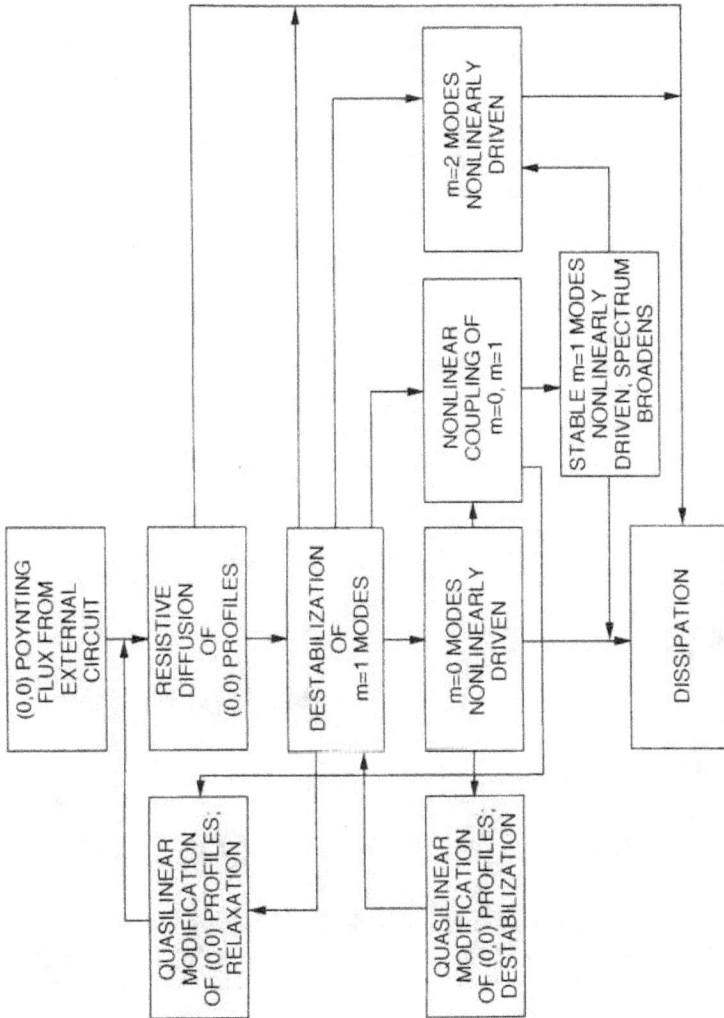

Figure 5-15. Schematic diagram of the RFP dynamo, indicating important processes.

details of the fluctuation spectrum. While their amplitude is individually low, they are all resonant at the field reversal surface $q = 0$, and can have a strong cumulative effect. In particular, they locally flatten the parallel current μ at the reversal surface (see Figure 5-9). The resultant steepening of the μ profile in the interior causes a second set of $m = 1$ modes, resonant further from the axis than the original dynamo modes, to be quasilinearly destabilized. The $m = 0$ spectrum becomes peaked at $n \approx 2\text{-}3$, corresponding to the nonlinear interaction between the two dominant parts of the $m = 1$ spectrum. The precise values of $|n|$ associated with the two peaks shifts in time as the mean profiles are quasilinearly altered near the axis. The quasilinear destabilization of $m = 1$ by $m = 0$ is as strong an effect as the original $m = 1$ destabilization by transport processes.

This picture has been obtained by numerical simulation of the force-free resistive MHD equations [Ho and Craddock, 1991] in which a detailed accounting was kept of the instantaneous power flux into and out of each Fourier mode (m,n). For example, the instantaneous power flux into the mode (m,n) can be written as

$$\Omega_{m,n} = \Omega_{q_{m,n}} + \Omega_{N_{m,n}} + \Omega_{D_{m,n}} , \qquad (5.18)$$

where

$$\Omega_{q_{m,n}} = \int \left[\mathbf{E}_{f_{m,n}} \cdot \mathbf{J}_0 - \langle \mathbf{j}_{m,n} \times \mathbf{b}_{m,n} \rangle \cdot \mathbf{V}_0 \right] d^3\mathbf{r} \qquad (5.19a)$$

is the quasilinear power flux,

$$\Omega_{N_{m,n}} = \int \left[(\mathbf{j} \times \mathbf{b})_{m,n} \cdot \mathbf{v}_{m,n} + (\mathbf{v} \times \mathbf{b})_{m,n} \cdot \mathbf{j}_{m,n} \right] d^3\mathbf{r} \qquad (5.19b)$$

is the nonlinear power flux, and

$$\Omega_{D_{m,n}} = \int \left[\nu \mathbf{v}_{m,n} \cdot \nabla^2 \mathbf{v}_{m,n} - \eta \mathbf{j}_{m,n}^2 \right] d^3\mathbf{r} \qquad (5.19c)$$

is the dissipation power flux; $\Omega_{qm,n}$ is the rate of energy transfer between the mode and the mean field (a similar expression, but with opposite sign, applies to the mean field), $\Omega_{Nm,n}$ is the rate of energy transfer directly between the mode (m,n) and all other modes, and $\Omega_{Dm,n}$ is the rate at which energy is removed from the mode by dissipative processes. Here, $\langle .. \rangle$ and $(..)_0$ both represent the mean value, and $\mathbf{E}_f$ is given by Eq. (5.12). By considering the energy in the mode (m,n),

$$E_{m,n} = \frac{1}{2} \int \left[v_{m,n}^2 + b_{m,n}^2 \right] d^3\mathbf{r} , \tag{5.20}$$

it is possible to define growth rates for quasilinear (γ_q), nonlinear (γ_N), and dissipative (γ_D) processes:

$$\gamma_{q_{m,n}} = \Omega_{q_{m,n}} / 2E_{m,n} , \tag{5.21a}$$

$$\gamma_{N_{m,n}} = \Omega_{N_{m,n}} / 2E_{m,n} , \tag{5.21b}$$

$$\gamma_{D_{m,n}} = \Omega_{D_{m,n}} / 2E_{m,n} . \tag{5.21c}$$

The total instantaneous growth rate is $\gamma = \gamma_q + \gamma_N + \gamma_D$. When instantaneous growth rates and power fluxes are computed for all relevant modes, a picture of power flow through Fourier space can be constructed. When combined with linear stability analysis, this can be used to elucidate the role of various modes in the dynamo process.

As an example of the use of Eqs. (5.19-5.21), we first analyze the growth and saturation of a single dynamo mode using the force-free model. This calculation is similar to the dynamical determination of helical ohmic states. As this system contains only a single mode (in addition to the mean fields), only quasilinear effects are possible. The initial mean state is non-reversed, and is unstable to a helical mode with mode numbers $m = 1$, $n = -5$; the Lundquist number is $S = 6 \times 10^3$. The evolution of the magnetic energy in the radial magnetic field, $W_{mr} = (1/2) \int B_r^2 \, d^3\mathbf{r}$, is shown in Figure 5-16a. After an initial linear growth phase, the $(1,-5)$ mode saturates after approximately 100 Alfvén times; a helical ohmic state is achieved. In Figure 5-16b we plot the corresponding time evolution of the instantaneous growth rates γ_q, γ_D, and $\gamma = \gamma_q + \gamma_D$ (recall that $\gamma_N = 0$ for this case). The dissipation rate is unchanged throughout the calculation, so that saturation is due entirely to the reduction in γ_q. The mode thus stabilized itself by modifying the mean fields. In the final steady state, the slightly positive quasilinear growth rate just balances the negative dissipation growth rate, thus illustrating the balance between instability and dissipation (transport) inherent in the RFP dynamo. In this simple case, this balance determines the saturation amplitude of the $(1,-5)$ mode. In Figure 5-16b we also display the *linear* growth and damping rates γ^* and γ_D^* obtained from a linear stability analysis of the final mean field profiles; note that they are exactly equal to the instantaneous growth rates obtained from Eqs. (5.19-5.21). Significant differences between γ_q and γ^* are indicative of strong nonlinear effects. Thus the helical Ohmic state is a quasilinear process.

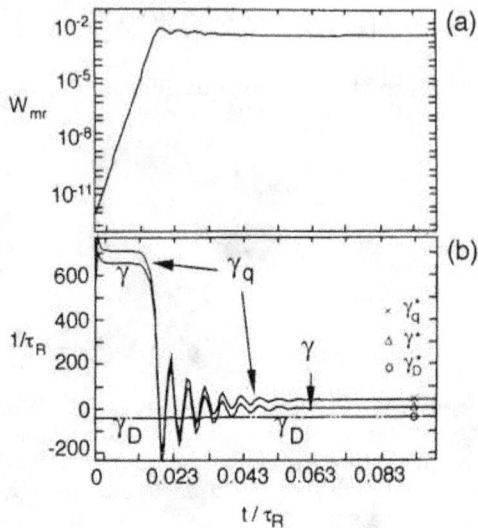

Figure 5-16. Quasilinear growth and saturation of the (1,–5) mode in an example case containing only one mode: **a.** radial magnetic energy, $W_{mr} = (1/2) \int B_r^2 d_r^3, r$, vs. time; **b.** quasilinear growth rate, γ_q, dissipation growth rate, γ_D and total growth rate ($\gamma = \gamma_q + \gamma_D$) vs. time. Without nonlinear mode coupling, γ, γ_q, and γ_D are the same as the growth rates obtained from a linearized calculation (γ^*, γ_q^*, and γ_D^*) where the mean field is frozen in time [*Ho and Craddock*, 1991].

As a more interesting situation, consider the case of a highly damped, low aspect ratio, force-free simulation with few interacting modes (only $m = 1$ and $m = 0$ modes, with $-10 \leq n \leq 10$, are included). The aspect ratio is 1, and the Lundquist number is $S = 10^3$. The dominant $m = 1$ modes in this system have n in the range -2 to -5. By slowly increasing the pinch parameter Θ, it is possible to isolate a single relaxation event characteristic of RFP sawtooth oscillations [*Watt and Nebel*, 1983; *Hutchinson et al.*, 1984; *Scime et al.*, 1992]. The time evolution of the pinch parameter Θ and the field reversal parameter F during this event are shown in Figure 5-17a. In Figure 5-17b we plot the corresponding time evolution of the radial magnetic energy in the $m = 1$, $n = -2$ to -5 modes, and in all the modes with $m = 0$. Prior to the relaxation event, the pinch existed in a helical Ohmic state dominated by the (1,–2) mode. After the event, the field reversal has increased (F has become more negative), the amplitude of the $m = 1$ modes has decreased slightly, and the amplitude of the $m = 0$ modes has increased significantly. The increase in field reversal signifies that an enhanced dynamo is present in the final state.

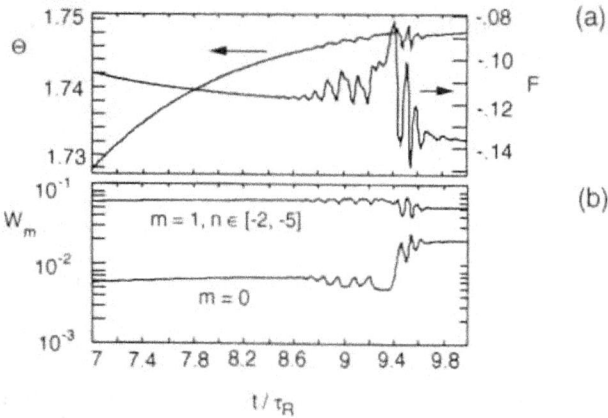

Figure 5-17. A single relaxation event: a. field reversal parameter F and pinch parameter Θ vs. time; b. magnetic energies W_m for all $m = 0$ modes, and $m = 1$, $n \in [-2,-5]$ modes vs. time. This is an example of a highly damped simple system containing few $(m \in [0,1], n \in [-10,10])$ interacting modes [*Ho and Craddock*, 1991].

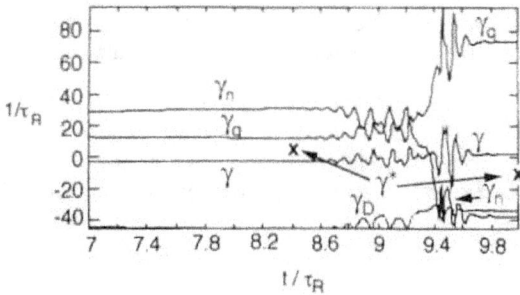

Figure 5-18. Time history of the total growth rate, γ, and its quasilinear (γ_q), nonlinear (γ_n), and dissipation (γ_D) for the $(1,-2)$ mode for the case shown in Figure 5-17. Growth rates from a linearized system (γ^*) are also evaluated a two different times [*Ho and Craddock*, 1991].

As a result of these calculations, Ho and Craddock were able to identify a three step nonlinear process that occurs during the relaxation event. The *first step* is the transfer of energy from the mean poloidal field to the $m = 1$ modes. In Figure 5-18 we plot the time evolution of the quasilinear, nonlinear, and dissipation growth rates [Eqs. (5.21(a-c))] for the $m = 1$, $n = -2$ mode during the relaxation event. In Figure 5-18 the linear growth rate γ^* before and after relaxation has also been identified; the $(1,-2)$ mode actually changes from being slightly linearly unstable to strongly linearly stable despite the increase in γ_q and mode amplitude. This is evidence of a strongly nonlinear process; linear stability provides no indication of the behavior of a particular mode. The $m = 0$ modes are clearly crucial here, since there are no $m = 2$ modes present in this example: all mode coupling involving $m = 1$ modes must first couple through $m = 0$.

The *second step* in the nonlinear dynamo is a nonlinear transfer of energy from $m = 1$ modes resonant near $r = 0$ (e.g., $m = 1, n = -2$) to $m = 1$ modes resonant nearer the reversal surface (e.g., $m = 1, n = -3$ to -5). From Figure 5-18 we note that after the relaxation event $\gamma_{q1,-2} \gg -\gamma_{D1,-2}$, so that the quasilinear power flux from the poloidal field is much more than is required to sustain the $(1,-2)$ mode. The excess energy flows out of the mode through nonlinear mode coupling ($\gamma_{N1,-2} < 0$). (As explained above, this power must flow *through* the $m = 0$ modes.) In Figure 5-19 we plot the various growth rates versus time for the $m = 1, n = -4$ mode. After relaxation, this mode is sustained largely by power influx through mode coupling ($\gamma_{N1,-4} > \gamma_{q1,-4}$), whereas prior to relaxation the mode relied mainly on the mean field as a power source (i.e., $\gamma_{q1,-4} \gg \gamma_{N1,-4}$).

The *third step* of the nonlinear dynamo is the transfer of energy from the $m = 1$ modes resonant near the reversal surface ($-5 \le n \le -3$) to the mean toroidal field. It is this transfer that sustains field reversal.

It is interesting to note that the $m = 0$ modes are *linearly stable* throughout the relaxation process. In fact, they act primarily as a drain of energy from the mean toroidal field, and thus have an *antidynamo* character. Nonetheless, they clearly play a significant role in determining the detailed

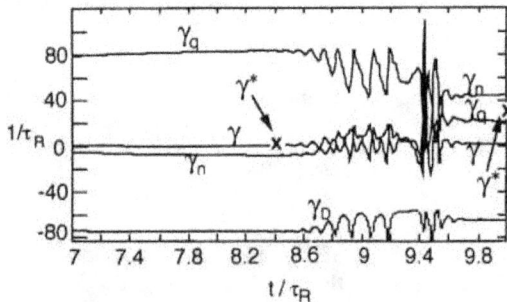

Figure 5-19. Time histories for the growth rates γ, γ_q, γ_n, and γ_D for the $(1,-4)$ mode for the case shown in Figures 5-17 and 5-18. The linear growth rate γ^* is also evaluated at two different times [*Ho and Craddock*, 1991].

properties of the final state. Because of their linear stability, their presence at all is the result of *nonlinear mode coupling* between the $m = 1$ modes, yet their amplitude is determined by quasilinear interaction with the mean fields. The $m = 1$ modes thus play a nonlinear, catalytic role in determining the presence of the $m = 0$ modes, and the quasilinear profile modification resulting from the $m = 0$ modes affects the $m = 1$ dynamics. Thus *a quasilinear picture alone is insufficient to describe relaxation in the RFP.*

By considering cases with larger aspect ratio, more modes, and less dissipation, it is possible to produce quasiperiodic oscillations. The time history of F, and the magnetic energy in the $m = 0$ and $m = 1$ modes, for a typical case is shown in Figure 5-20. The same nonlinear dynamics described above are important. (Since coupling to $m = 2$ was included in these cases, an extra source of dissipation was present.) After a relaxation event, ordinary resistive diffusion alters the mean profiles in such a way as to drive the $m = 1$ modes in the interior, allowing repeated relaxation to occur.

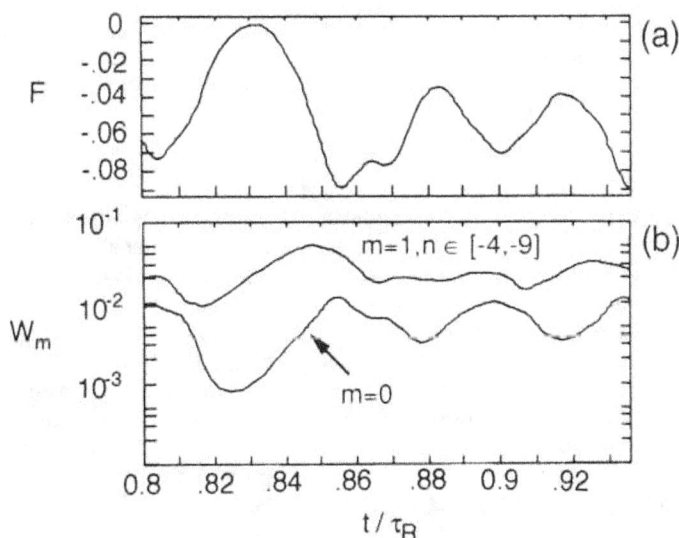

Figure 5-20. Quasiperiodic oscillations for a case with Lundquist number $S = 6 \times 10^3$ and aspect ration $R/a = 2.5$: a. field reversal parameter F vs. time; b. magnetic energies for all $m = 0$ modes and dominant $m = 1$ modes vs. time [*Ho and Craddock*, 1991].

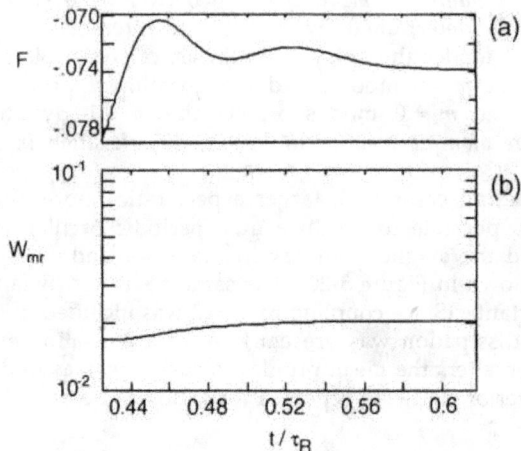

Figure 5-21. Time history of **a.** the field reversal parameter, F, and **b.** total radial magnetic energy for a test case with $m = 0$ modes artificially removed. All other parameters are identical to the case shown in Figure 5-20 [*Ho and Craddock*, 1991].

The important role played by the $m = 0$ modes in determining the observed dynamical oscillations is illustrated in Figure 5-21, where we plot the time history of F and the modal magnetic energy for a case in which the $m = 0$ modes were artificially removed from the calculation. While the mean toroidal field remains reversed, the pinch now settles into a steady, nonoscillating state that does not resemble experimental observations: the physical state of the system has been entirely changed. By removal of the $m = 0$ modes, nonlinear coupling and spectral broadening have been eliminated, and the efficient transfer of energy from the poloidal magnetic field in the interior to the toroidal magnetic field near the edge has been inhibited. The simulation now resembles a highly damped system, even though the physical dissipation has not been altered.

5.4 Summary

In the RFP magnetoplasma configuration, plasma relaxation produces toroidal flux from poloidal flux. The poloidal flux must be continuously provided from an external source (such as an electric circuit); the regeneration of the reversed-field configuration proceeds only as long as this source remains active. This differs significantly from the classical dynamo picture, in which the magnetic configuration is sustained by an externally driven velocity field.

Relaxation in the RFP occurs as the result of the nonlinear evolution of resistive kink modes with poloidal mode number $m = 1$. These modes become unstable as a result of resistive diffusion that drives the profiles away from the preferred relaxed state. Their nonlinear evolution rearranges the electric current distribution in such a way as to restore the relaxed profiles. Sustained, relaxed steady states can result from the quasilinear evolution of a single mode. During this relaxation process, the ratio of total magnetic energy to total magnetic helicity in minimized, as hypothesized by Taylor.

When several resistive kink modes are unstable simultaneously, their nonlinear interaction results in a richly dynamic evolution of the sort observed in experiments. In the RFP, the nonlinear relaxation dynamics then proceed in three steps. The first step is a transfer of energy from the mean poloidal magnetic field to the unstable $m = 1$ resistive kink modes active near the core of the plasma. These modes nonlinearly generate other modes with poloidal mode number $m = 0$. The second step is a transfer of energy from these modes to other $m = 1$ modes active near the reversal surface. This energy flows through the $m = 0$ modes, and is purely a nonlinear process. (When the $m = 0$ modes are artificially removed from these calculations the rich dynamical character ceases.) The third step is the quasilinear transfer of energy from the $m = 1$ modes active near the reversal surface to the mean toroidal field. In this way, the mean poloidal flux provided by the external circuit is converted into the mean toroidal flux required to sustain the reversed field configuration. The overall relaxation dynamics are nonlinear; a quasilinear picture alone is insufficient.

Magnetic fluctuations due to long wavelength MHD instabilities can thus account for many of the observed properties of plasma relaxation, at least in the RFP. It turns out that these same relaxation dynamics can self-consistently account for the ubiquitous presence of anomalous loop voltage in experimental discharges, sawtooth-like oscillations and their dependence on externally controllable parameters, the observed phase locking of magnetic fluctuations, and the operating characteristics of plasma discharges with resistive walls and limiters. An interpretation of these results in terms of global helicity balance can also be given. As a result, an experimentally testable concept for controlling the magnetic fluctuation level and sustaining current in the RFP can be proposed. These issues are all discussed in more detail in Chapter 6.

CHAPTER 6

PRACTICAL ISSUES RELATED TO RELAXATION

In Chapter 5 the dynamical mechanism responsible for plasma relaxation was elucidated. It was demonstrated that the nonlinear evolution and interaction of long wavelength resistive MHD instabilities can produce magnetoplasma configurations in close agreement with those predicted by Taylor's variational theory. These instabilities are triggered by the action of resistive diffusion, which is continually driving the system away from its preferred relaxed state. Thus relaxing plasmas can be expected to exhibit some level of intrinsic magnetic and velocity fluctuations.

In this chapter, and the next, we consider some of the consequences that these intrinsic fluctuations may have for the operation of plasma experiments, and for the transport of energy through the plasma. These issues are particularly important in the fusion energy program, where the goal is to confine a hot plasma for a long enough time that net fusion energy can be produced. However, as we have throughout, we will continue to concentrate on physics, rather than technology, issues. This chapter deals with the electromagnetic consequences of plasma relaxation. Chapter 7 will address the relationship between plasma relaxation and transport.

Again, we will consider the Reversed-field Pinch as a paradigm for the study of relaxation and its consequences. Reversed-field Pinch plasmas characteristically require an anomalously large loop voltage to drive a given amount of toroidal current. In the past, this voltage anomaly has been interpreted in terms of an enhancement, of unknown origin, to the plasma resistivity. Instead, we will see that an increase in loop voltage is a natural consequence of the magnetic fluctuations that are responsible for plasma relaxation. These fluctuations can also account for the observed increase in loop voltage and fluctuation level when experiments are operated with a resistive, rather than perfectly conducting, outer boundary. An examination of global helicity balance in numerical studies of this type of RFP operation then leads to a conceptual description of experimental means of controlling the dynamo fluctuations, and hence reducing the loop voltage.

6.1 Anomalous Loop Voltage

Perfectly Conducting Outer Boundary

Experimentally, RFP discharges exhibit an anomalous plasma resistance that results in larger loop voltage requirements than would be expected from considerations of Coulomb collisions alone. Recall that the parallel component of the electric field arising from dynamo fluctuations, $E_f = - \langle v \times b \rangle$, *suppresses* parallel current on axis and *drives* parallel current at the edge [Ho et al., 1989]. This was discussed in Section 5.2. The result of the current suppression on axis is an anomalous loop voltage, or plasma resistance, larger than that obtained from simple Ohmic considerations with classical (Coulomb) resistivity. For the case of a sustained RFP with a perfectly conducting outer boundary shown in Figure 5-7, the voltage is about 43 percent anomalous. (In the absence of fluctuating magnetic and velocity fields, the applied electric field is simply given by the parallel component of ηJ.)

It is to be emphasized that this anomalous loop voltage, or effective plasma resistance, arises from the MHD fluctuations alone; the actual plasma resistivity η may still be classical. The effect of the dynamo fluctuations is to convert poloidal flux into toroidal flux. Thus the external circuit must not only supply the poloidal flux to maintain the desired mean current, but must feed the dynamo as well. The more active the dynamo, the more excess flux must be supplied. This excess rate of poloidal flux input appears as anomalous loop voltage.

Operation with Resistive Walls and Limiters

Historically, RFP theories and experiments have included a highly conducting boundary near the plasma surface. As discussed in Section 2.3, it is well known that such a boundary condition is necessary for linear MHD stability [Robinson, 1978]. A conducting outer boundary also plays an integral role in the Taylor relaxation theory, which requires the boundary condition $\hat{n} \cdot B = 0$ (see Section 3.1). However, in order to optimize plasma position control and reduce magnetic field errors, thin (resistive) boundaries may be required in long pulse experiments [Malesani, 1988]. More fundamentally, there is no conducting wall in long-pulse of steady state operation: all shells become resistive on a long enough time scale. It is therefore of interest to study the effect of such a nonideal boundary on RFP operation in general, and on the RFP dynamo in particular.

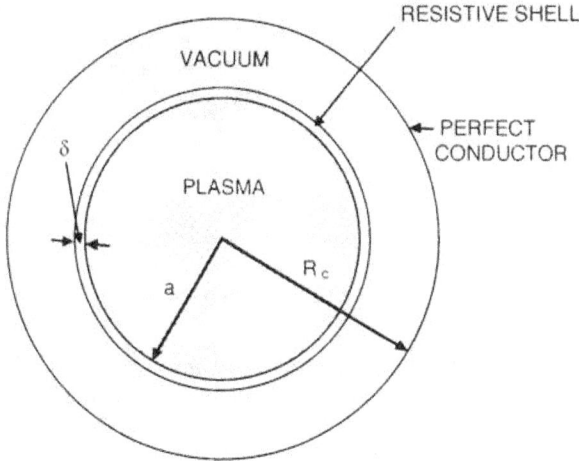

Figure 6-1. Sketch of geometry for a cylindrical RFP with a resistive shell.

At a perfectly conducting wall, the normal component of the magnetic field and the fluctuating tangential electric field must vanish. (The mean tangential electric field is the specified applied voltage.) When the plasma boundary is not perfectly conducting, boundary conditions appropriate to a thin, resistive shell must be derived. We envision a case in which the plasma is bounded by a resistive shell at $r = a$. The magnetic field in the plasma will be denoted as $\mathbf{B}_p$. A perfectly conducting boundary may exist at a larger radius $r = R_c$. The region $a < r < R_c$ contains vacuum, in which the magnetic field is given by $\mathbf{B}_v = \nabla\phi$, where $\nabla^2\phi = 0$. The geometry is sketched in Figure 6-1. At the outer conducting shell $r = R_0$, the normal component of $\mathbf{B}_v$ must vanish, so $\hat{n} \cdot \nabla\phi = 0$ there. A boundary condition to determine the evolution of the fields at $r = a$ can be obtained by applying the proper jump conditions on the magnetic field at the resistive shell.

We assume that the shell has a thickness δ that is much less than any other macroscopic length scale, and that it can be characterized by a resistivity η_s. The jump conditions are

$$\hat{n} \times [\mathbf{B}] = \mathbf{K} \,, \tag{6.1a}$$

$$\hat{n} \cdot [\mathbf{B}] = 0 \,, \tag{6.1b}$$

where $\hat{n}$ is the outward drawn normal from the plasma, K is a surface current density, and $[B]$ is the jump in B across the thin shell at $r = a$, $[B] = B_v(a) - B_p(a)$. In cylindrical geometry, $\hat{n}$ is in the radial direction. We also write $K = J_s\delta$, where $J_s = E(a)/\eta_s$ is the current density flowing in the shell, and $E(a)$ is the electric field at $r = a$. Then, using $B_p = \nabla \times A$ and $E = -\partial A/\partial t$, Eq. (6.1) can be written as

$$\frac{\tau_s}{a}\left.\frac{\partial A}{\partial t}\right|_a = \hat{n} \times (\nabla \times A - \nabla\phi)\big|_a , \tag{6.2a}$$

$$\hat{n}\cdot\nabla\phi\big|_a = \hat{n}\cdot\nabla\times A\big|_a , \tag{6.2b}$$

where $\tau_s = a\delta/\eta_s$ is the time constant for the resistive shell. Equations (6.2a,b) are the matching conditions that connect the vacuum and plasma solutions at $r = a$.

The magnetic potential ϕ contains two constants of integration. There are thus four quantities to be determined: the two constants of integration, and the two tangential components of the vector potential A at $r = a$. (The radial component of A is determined completely by conditions in the plasma [*Schnack et al.*, 1987]). We have four equations to determine these quantities: Eq. (6.2b), the two tangential components of Eq. (6.2a), and the condition $\hat{n}\cdot\nabla\phi = 0$ at $r = R_c$.

In cylindrical geometry, Eqs. (6.2) take a particularly simple form. The solution can be Fourier decomposed into modes with poloidal mode number m and toroidal mode number n. The vacuum solution is given by

$$\phi_{m,k}(r) = \alpha_{m,k}I_m(x_k) + \beta_{m,k}K_m(x_k) , \tag{6.3}$$

where $k = na/R$, $x = kr$, α and β are the constants of integration, and I_m and K_m are modified Bessel functions. There are four unknowns to be determined for each mode (m,k): α, β, A_θ and A_z. The condition $B_r(R_c) = 0$ determines one of the constants α or β; Eq. (6.2b) determines the other, and allows $\phi(a)$ to be expressed in terms of the tangential components of A. In cylindrical geometry, $\phi(a)$ enters the tangential components of Eq. (6.2a) only algebraically, and can thus be eliminated. The resulting equations are

$$-\tau_s\frac{\partial A_\theta}{\partial t} = \frac{\partial}{\partial r}(rA_\theta) + imA_r + ikaC\left(-\frac{im}{a}A_z + ikA_\theta\right) , \tag{6.4a}$$

$$-\tau_s\frac{\partial A_z}{\partial t} = a\frac{\partial A_z}{\partial r} + ikaA_r - imC\left(-\frac{im}{a}A_z + ikA_\theta\right) . \tag{6.4b}$$

Again, A_r is to be considered as determined by conditions within the plasma. These equations describe the evolution of A for each mode at $r = a$. The constant C couples the vacuum and plasma solutions, and is given by

$$C = \frac{I'_m(kR_c)\, K_m(ka) - K'_m(kR_c)\, I_m(ka)}{k\left[I'_m(kR_c)\, K'_m(ka) - K'_m(kR_c)\, I'_m(ka)\right]}\,, \qquad (6.5)$$

where $(...)'$ denotes differentiation with respect to x.

We have seen in Section 2.3 that, in the RFP, MHD stability is greatly influenced by the position of a conducting boundary where the linear eigenfunction must vanish. This is a result of requiring $\hat{n} \cdot B = 0$ there. When the outer boundary is resistive, this condition is relaxed and these modes are destabilized. Recall that the RFP dynamo is driven by $m = 1$ modes resonant near the magnetic axis. Linear theory thus predicts that these modes are further destabilized by a resistive outer boundary, as are both $m = 0$ internal modes and $m = 1$ external kink modes [Gimblett, 1986; Ho and Prager, 1988]. The latter have a pitch opposite to that of the dynamo modes and are nonresonant from below, i.e., beyond the outer boundary. The nonlinear evolution of these modes has been studied [Ho et al., 1989; Schnack and Ortolani, 1990; Ho and Prager, 1991] by first obtaining a sustained RFP discharge with an active dynamo in the presence of a perfectly conducting wall, and then introducing the boundary condition at $r = a$ that models a resistive shell with skin time τ_s. A perfectly conducting boundary may exist at some larger radius $r = R_c > a$, with vacuum in the region $a < r < R_c$. A variety of conditions can then be modeled: finite τ_s and infinite R_c effectively models a resistive shell; small τ_s and finite R_c models operation with a limiter. The effect of the resistive outer boundary on the further nonlinear evolution of the discharge can then be studied in detail.

It is found [Ho et al., 1989; Ho and Prager, 1991] that the destabilization of the internal dynamo kink modes in the presence of a resistive outer boundary causes an increase in the fluctuation level and a corresponding increase in the dynamo electric field produced by these fluctuations. For the conducting wall case, this field is plotted in Figure 5-7. In Figure 6-2 we plot the same components of the parallel electric field in the presence of a resistive shell, with $\tau_s = 0.1\tau_R$ and $R_c \to \infty$. All other parameters are the same as in Figure 5-7; in particular, the toroidal current has been held constant. We see that E_f is now about seven times the resistive term on axis. To balance the current suppression of E_f the loop voltage is now highly anomalous.

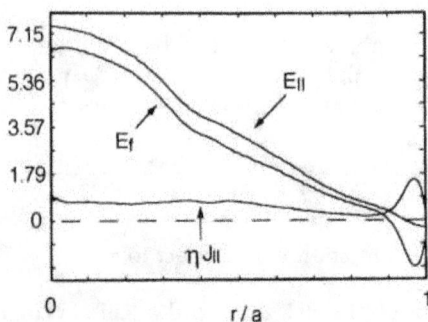

Figure 6-2. Average parallel electric fields vs. radius during the sustainment phase of a three-dimensional numerical simulation of an RFP discharge with a resistive outer boundary. Ohmic, fluctuating (dynamo), and total fields are shown. Fields are normalized to the Ohmic electric field on axis. In the absence of fluctuations, the field would be entirely Ohmic; the fluctuations result in a total electric field that is more than 7 times the Ohmic value on axis. Compare with Figure 5-7, where the same case with a perfectly conducting shell is shown.

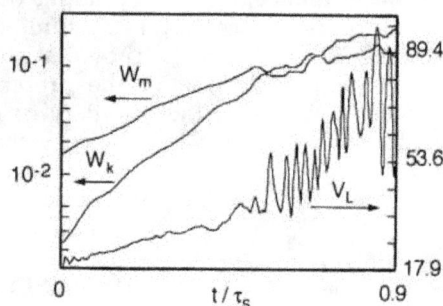

Figure 6-3. Constant Θ evolution (initialized with a close-fitting conducting wall steady state) with a thin (resistive) shell. The total radial magnetic energy, W_m, the kinetic energy (measure in the same units as W_m), and the loop voltage V_L, are plotted vs. time (measured in units of the resistive shell time constant).

In Figure 6-3 we plot the loop voltage V_L, the fluctuating magnetic energy W_M, and the fluctuating kinetic energy W_K, as functions of time (normalized to the skin time of the shell) after the introduction of the resistive shell boundary condition for a case with $\Theta = 1.59$. We see that the loop voltage increases dramatically with time, reaching values of three to five times that required to sustain the current with a close fitting conducting shell. The fluctuations also increase as the dynamo becomes more active, with W_M increasing by one order of magnitude, and W_K increasing by two orders of magnitude.   This relatively larger increase in kinetic energy with respect to fluctuating magnetic energy has been found to be characteristic of resistive shell operation.

In some experiments small strips of material protrude into the plasma from the shell in order to limit the interaction between the plasma and the shell [Alper et al., 1989a,b].   These limiters effectively introduce a vacuum region between the plasma and the conducting outer boundary.   The plasma-vacuum boundary can be modeled as a resistive shell with a very short time constant: the boundary is impermeable to plasma flow but transparent to magnetic flux.   The effect of limiters on experimental operation can therefore be studied by setting $\tau_s = 0.01\tau_R$, and gradually increasing R_c as a function of time. The results of such a study are shown in Figure 6-4, where we plot the loop voltage as a function of vacuum region thickness $(R_c - a)/a$ for the case $\Theta = 1.59$. We see that V_L rises with R_c, increasing dramatically beyond $R_c = 1.33a$, at which point oscillations in V_L begin.   (Recall that the toroidal current is held constant throughout the wall expansion process; the oscillations at large R_c represent the attempts of the external circuit to hold the current precisely constant in the presence of increasingly large plasma oscillations and inductance changes.   The same phenomena is responsible for the large loop voltage fluctuations that appear in Figure 6-3.) At high Θ values the rise in V_L is more dramatic. At $\Theta = 1.73$, V_L rises by 50 percent as R_c increases from a to $1.05a$.

The results described above are in substantial agreement with RFP experiments with limiters and resistive shells where increases in loop voltage and fluctuation levels have been observed [Tamano et al., 1987; Alper et al., 1989a,b].   The dynamo modes are destabilized by the resistive shell; the resulting larger amplitude fluctuations create a stronger dynamo: poloidal flux is converted into toroidal flux at a faster rate.   This extra poloidal flux must be supplied by the external circuit to keep the toroidal current fixed.   But

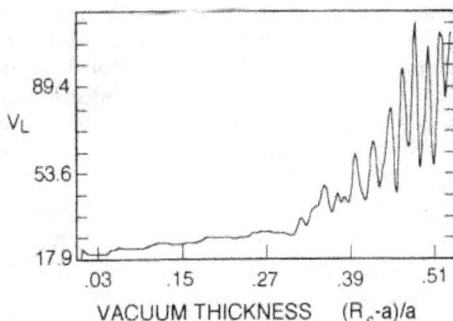

Figure 6-4. Loop voltage, V_L, vs. vacuum region thickness for a case with $\Theta = 1.59$ and $\tau_S = 0.01\tau_R$.

the rate at which poloidal flux is supplied by the external circuit is just the toroidal voltage; hence, the loop voltage increases. Thus, for the case of limiter operation, the fluctuations and loop voltage increase with vacuum region thickness.

In all cases, even with a perfectly conducting boundary, fluctuations increase with Θ because the field reversal surface then becomes relatively farther from the wall, and there is more free energy in the mean currents to drive instabilities. Also, the $m = 0$ modes that are essential for the nonlinear dynamo (see Section 5.3) are resonant further from the wall, and can be driven to larger amplitude. External kink modes have been found at high Θ, where their effect is to further increase the loop voltage and, in extreme cases, terminate the discharge [Schnack and Ortolani, 1990]; they do not appear when $\Theta < 1.6$, or at low S.

It has been found that the loop voltage and fluctuation increases described in this section are independent of the shell skin time τ_s, but rather inevitably appear after several shell times [Ho et al., 1989; Schnack and Ortolani, 1990].

Finally, among the more curious of RFP experimental results is the observation of the "slinky" mode. This mode was originally seen during resistive shell operation [Tamano et al., 1987], but similar features have also been observed in the presence of a perfect conductor [Prager, 1990]. This mode is characterized by the appearance of a toroidally localized disturbance that grows on a time scale determined by the resistive shell time constant, a flattening or hollowing of the parallel current profile, an increase in $q(0)$, a shift of the $m = 1$ spectrum to longer wavelengths, and a phase locking of the internally resonant $m = 1$ (dynamo) modes [Tamano et al., 1987].

Similar feature have been observed in nonlinear MHD simulations of RFPs [Schnack and Ortolani, 1990; Kusano et al., 1991]. An example is shown in Figure 6-5a,b, where the radial magnetic field and axial current density at the wall are plotted as functions of axial distance (toroidal angle) at a time well into a simulation with a resistive shell with $R/a = 4$ [Schnack and Ortolani, 1990]. Note the toroidally localized disturbance in each signal. Figure 6-6 shows the phases (defined, for example, as $\phi_{m,n} = \tan^{-1}[\mathrm{Im}(B_{rm,n})/\mathrm{Re}(B_{rm,n})]$) of internally resonant $m = 1$ modes at the wall as functions of time. The distinct phase locking (modes having equal phase) occurs only for the internally resonant dynamo modes with $-12 \leq n \leq -9$. Modes that are resonant near the reversal surface, externally resonant modes, and modes that are nonresonant from above do not exhibit this phase locking. This is in substantial agreement with experimental observations [Tamano et al., 1987].

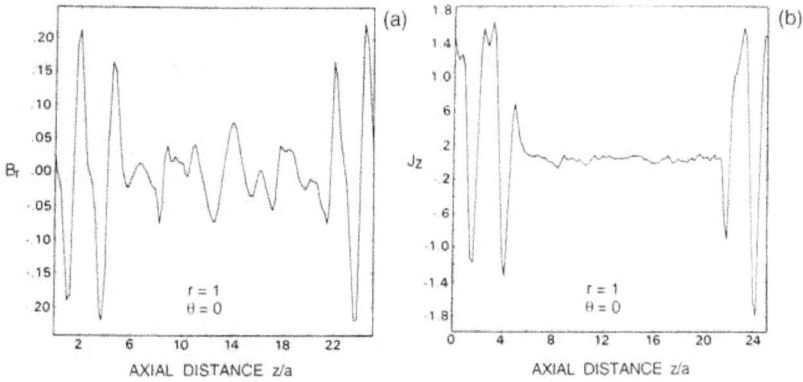

Figure 6-5. **a.** Radial magnetic field, B_r, and **b.** axial current density, J_z, vs. axial distance z/a, at the outer boundary at a given time during the numerical simulation of an RFP with a resistive shell. Note the axially localized disturbance.

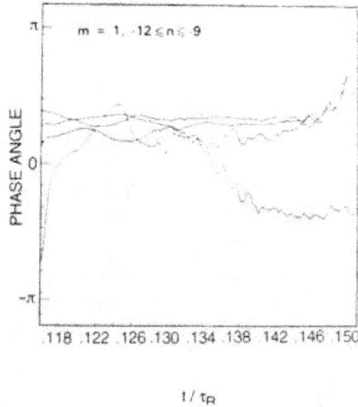

Figure 6-6. Phase $\phi_{m,n} = \tan^{-1}[\mathrm{Im}(B_{rm,n})/\mathrm{Re}(B_{rm,n})]$ vs. time for internally resonant modes. Note the phase locking that occurs between $t/\tau_R = 0.12$ and $t/\tau_R = 0.134$. The axially localized disturbance shown in Figure 6-5 occurs during this time interval.

In addition to the resistive shell configuration, cases with perfectly conducting boundaries have been examined [Schnack and Ortolani, 1990; Kusano et al., 1991], and a similar phase locking has been found. Thus the phase locking of resistive modes does not depend on the presence of a resistive shell, but is a more general property of RFP operation in particular, and of generic nonlinear mode coupling in the MHD model [Kusano et al., 1991].

Helicity Balance

It is instructive to interpret the results pertaining to operation with resistive boundaries in terms of conservation of magnetic helicity [*Ho et al.*, 1989]. However, the results must be interpreted carefully in the presence of a resistive boundary, since $\hat{n} \cdot \mathbf{B} \neq 0$ in the bounding surface (see Section 3.2). For our applications, we must also use the properly defined helicity [see Eq. (3.39)]. In this case, the time derivative of K is given by

$$\frac{dK}{dt} = -2 \int \eta \mathbf{J} \cdot \mathbf{B} dV + 2 \oint \phi \mathbf{B} \cdot \hat{n} dS + 2 \oint \mathbf{A} \cdot d\mathbf{l}_1 \oint \mathbf{E} \cdot d\mathbf{l}_2 + \oint \chi \frac{\partial \mathbf{B}}{\partial t} \cdot \hat{n} dS , \quad (6.6)$$

where ϕ is the electrostatic potential, χ is the gauge potential, and we have used $\hat{n} dS = d\mathbf{l}_1 \times d\mathbf{l}_2$. Recall that χ is a purely mathematical construction that can be changed arbitrarily; in this context it has no physical significance. Thus, dK/dt (as well as K itself) is physically ill defined, being explicitly dependent on χ. However, this ambiguity is removed in the steady state where $d/dt = 0$. In this situation *helicity balance* can be written as

$$\oint \mathbf{A} \cdot d\mathbf{l}_1 \oint \mathbf{E} \cdot d\mathbf{l}_2 = \int \eta \mathbf{J} \cdot \mathbf{B} dV - \oint \phi \mathbf{B} \cdot \hat{n} dS . \quad (6.7)$$

The left-hand-side of Eq. (6.7) is a measure of the voltage required to sustain the discharge in steady state. The first integral on the right-hand-side of Eq. (6.7) is the dissipation of helicity by resistive diffusion. The second integral on the right-hand-side of Eq. (6.7) is to be interpreted as the leakage of helicity through the boundary that is penetrated by magnetic field lines. Note that ϕ is the electrostatic potential, and has physical significance. [Moreover, Eq. (6.7) is unaffected by the addition of a constant to ϕ, since such a transformation leaves the last integral unaffected due to the vanishing of the average of $\mathbf{B} \cdot \hat{n}$]. In steady state, ϕ is computed directly from $\mathbf{E} = \nabla \phi = -\partial \mathbf{A} / \partial t$.

In toroidal or periodic cylindrical geometry, helicity balance can be written as

$$V_L \Phi_z = \int \eta \mathbf{J}_0 \cdot \mathbf{B}_0 dV + \int \eta \mathbf{j} \cdot \mathbf{b} dV - \int \phi \mathbf{b} \cdot d\mathbf{s} . \quad (6.8)$$

Here, V_L is the toroidal (axial) loop voltage, Φ_z is the toroidal (axial) flux, and we have separated $\mathbf{J}$ and $\mathbf{B}$ into parts that depend only on radius, with subscript $(..)_0$, and those that depend on all three spatial coordinates (i.e., the spatial fluctuations, denoted by lower case letters). The left-hand-side gives the loop voltage required to sustain the pinch. The first term on the right-hand-side measures helicity dissipation by the mean fields, the second

Table 6-1. Helicity balance for resistive shell RFP.

Θ, % vacuum	$V_L\Phi_z$	$\int \eta J_0 \cdot B_0 dV$	$\int \eta j \cdot b \, dV$	$-\int \phi b \cdot ds$
1.592, 0%	24.5	24.9	0.5	0
1.592, 15%	32.9	25.1	1.4	6.2
1.592, 45%	43.6	25.5	2.6	13.1
1.73, 0%	32.7	33.7	0.8	0
1.73, 5%	48.2	37	2.7	6.2

term measures helicity dissipation by the fluctuations (e.g., the dynamo modes), and the third term measures the helicity loss (or gain) at the wall.

The terms on the right-hand-side of Eq. (6.8) have been evaluated for the numerical simulations described above [Ho et al., 1989], and are tabulated in Table 6-1. These results are not in perfect steady state so that exact balance is not attained. However, the trend is clear. These results show that the mean field resistive helicity dissipation in the plasma increases negligibly with enhanced fluctuations at constant Θ. Instead, the increased helicity input ($V_L\Phi_z$) is lost through the surface term. The MHD model includes E_f as the mechanism causing the rise in V_L. The vacuum region is destabilizing, and the current flow is interrupted when the perturbed magnetic field enters the vacuum region, leading to surface helicity loss. Vacuum induced instability implies surface helicity leakage, and *vice versa*. The instability both enhances the helicity input by raising V_L (through E_f) and enhances the surface helicity loss.

6.2 Taming the Dynamo; An Application of the Theory

We have seen that the dynamo has both favorable and unfavorable consequences for the RFP. On one hand, it is responsible for the sustainment of the discharge for times much longer than that predicted by resistive diffusion; there would be no RFP without the dynamo. On the other hand, the magnetic fluctuations responsible for the dynamo cause the loop voltage to be anomalously large, thus increasing the power input required to produce and maintain the mean plasma profiles. They may therefore entirely prevent the operation of long pulse devices with resistive shells; all shells become resistive on a long enough time scale. It would be desirable to operate the RFP without the dynamo, or at least have a method of externally controlling the dynamo fluctuations. In that way a proper balance between dynamo sustainment, transport, and loop voltage may be attainable. Such a mechanism [Ho, 1991] is suggested in this section.

The key ingredient in such a scheme is contained in Eq. (6.8). We have seen that the loop voltage, the dynamo fluctuations, and the surface helicity loss are self-consistently linked: dynamo fluctuations cause a suppression of current, an increase in V_L, and an increase in surface loss. Thus a change in any one term in Eq. (6.8) causes self-consistent changes in the others; however, only the surface loss term is externally controllable. It is thus proposed [Taylor and Turner, 1989; Fernandez et al., 1989; Ho, 1991] that, if the surface potential ϕ can be specified by external means in such a way as to decrease the surface losses, the fluctuation level and loop voltage may decrease as well. In the extreme case, if the sign of the surface term is changed so that helicity is actually injected into the discharge, perhaps the dynamo can be completely eliminated and the loop voltage reduced to a classical value.

In one such scheme, some of the mean field is diverted from the discharge by an external conductor, using a poloidal divertor, for example. This is required to obtain finite value for the normal flux at the wall. An example is sketched in Figure 6-7. Electrostatic potential can then be applied to external plates intersected by the diverted field, or, equivalently, electric current can be injected along the diverted field lines by electrodes. Thus the surface loss term in Eq. (6.8) is at the control of the experimenter.

The procedure described above, and sketched in Figure 6-7, has been simulated numerically [Ho, 1991]. In Figure 6-8 we plot the magnetic energy in the dynamo modes as a function of time after the beginning of helicity injection. (The $m = 1$, $n = 0$ mode is imposed by the poloidal divertor.) It is seen that the dynamo modes are completely eliminated while an interesting RFP state is sustained. The loop voltage also drops during this process. Thus, in principle, an RFP can be sustained indefinitely without magnetic fluctuations by DC helicity injection.

The procedure and results described in this section are speculative and technically simplistic, and hence are likely to be optimistic. Nonetheless, they represent a theoretical "proof of principle", and indicate that the fluctuation level and its experimental consequences, such as anomalously large loop voltage, may be externally controllable. Perhaps plasma relaxation can be induced, rather than simply accepted, in future experiments.

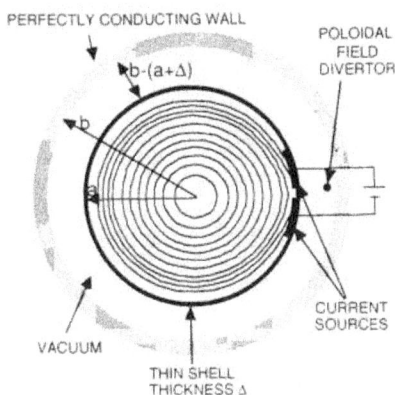

Figure 6-7. Typical RFP configuration with DC helicity injection through electrodes (labeled as current sources). The torus is modeled as a periodic cylinder [Ho, 1991].

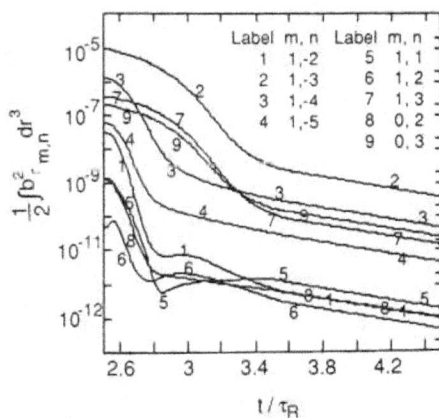

Figure 6-8. Radial magnetic energy vs. time for selected modes during DC helicity injection. The decrease in modal energy corresponds to an increase in injected helicity [Ho, 1991].

CHAPTER 7

RELAXATION AND THERMAL TRANSPORT

We have seen that magnetic fluctuations are a ubiquitous feature of RFP operation; they are, in fact, responsible for the sustainment of the discharge through the process of plasma relaxation. We have also seen that these same fluctuations can cause anomalously large toroidal loop voltage, and can adversely affect experimental operation with resistive outer boundaries. These are electromagnetic issues, i.e., they are describable without reference to thermodynamics or plasma thermal energy. We will now briefly address the effect that these fluctuations may have on plasma thermal energy and its transport.

By the term transport, we mean the conveyance of energy from one (relatively hot) part of the plasma to another (relatively cold) part. We consider transport of energy by conduction or convection. (Radiation can be an important energy loss mechanism, but does not contribute to transport since typical laboratory plasmas are optically thin.) The details of transport processes in magnetized plasmas are not well understood in general, and may be dependent on the specific geometry of the magnetoplasma system involved. However, virtually all experimental magnetized plasmas exhibit thermal transport that is much more rapid than that expected on the basis of classical mean free path considerations. The same is likely true of naturally occurring plasmas, although they are not so readily studied. This is clearly an important issue in the program to produce a fusion power reactor, where a hot plasma must be confined by a magnetic field for times sufficiently long for net fusion power to be produced. It is also an interesting unsolved physics problem.

The transport of thermal energy by conduction in a magnetized plasma is highly anisotropic, being very rapid in the direction parallel to the magnetic field, and relatively inefficient in any direction perpendicular to it. In an axisymmetric system with no magnetic fluctuations, the magnetic field lines form closed, nested flux surfaces. The transport of energy from the core of the plasma to the edge is then primarily determined by thermal conduction perpendicular to the magnetic field. We will find that magnetic fluctuations that are responsible for plasma relaxation can lead to a mixing of the magnetic field lines in the plasma in such a way that a single field line no longer traces out a flux surface, but rather can fill a large fraction of the plasma volume. The rapid transport of thermal energy along these stochastic field lines may result in an anomalously rapid loss of energy from the core of the plasma to the edge compared to that expected from perpendicular thermal conduction alone. At least in the Reversed-field Pinch, this process may contribute significantly to the experimentally measured confinement properties.

In this short chapter we describe extensions of the MHD calculations reported in Chapter 5 that allow the effect of plasma relaxation on energy transport to be evaluated quantitatively. This requires that finite pressure effects be added to the momentum equation, and that an energy equation with Ohmic heating and anisotropic thermal conduction be solved. All local transport *coefficients* that appear in this equation, such as electrical resistivity and thermal conductivity, are taken to be classical (i.e., arising from Coulomb collisions) [*Braginskii*, 1965]. However, the *global* rate of energy transport from the plasma to the wall that results from these calculations may differ significantly from that expected on the basis of these coefficients alone: enhanced radial transport may be induced by the fluctuations that are responsible for relaxation. This will allow dynamical relaxed states with finite plasma pressure to be determined. Overall values of β and energy confinement time can then be found and compared with experiment.

7.1 A Model for Sawtooth Oscillations in the RFP

Experimental Observations

Reversed-field Pinch discharges are observed to exhibit large amplitude oscillations in the soft x-ray flux from the core of the discharge [*Watt and Nebel*, 1983; *Wurden*, 1984]. These soft x-rays are a measure of the temperature in the central region of the plasma. The oscillations exhibit an asymmetric character, and are called sawtooth oscillations. Typical experimental observations are shown in Figure 4-18, which shows the time history of several quantities during experimental operation. These oscillations have two distinct phases: a slow rise phase, during which the core of the plasma is slowly heated; and, a rapid crash phase during which the plasma core is rapidly cooled. The sawtooth-like character of the oscillations comes from the discrepancy in time scales associated with the two phases. The fall of the central temperature suggests that some rapid energy transport process is active during the sawtooth crash phase.

The rise phase of the sawtooth oscillation is quantitatively describable in terms of plasma heating due to resistive diffusion, as has been verified by one-dimensional transport calculations [*Werley et al.* 1985]. This is essentially the global thermal instability described in Section 5.2. During this phase we then expect the magnetic field profiles to be driven away from their preferred relaxed state, and global resistive kink modes to be driven unstable. This has been verified by monitoring the linear growth rate of these modes during transport simulations of the rise phase of the sawtooth oscillation; these growth rates reach large values by the time of the onset of the sawtooth crash phase.

The sawtooth crash is associated with large $m = 0$ and $m = 1$ magnetic fluctuations, and with the generation of mean toroidal flux. Experimental results are shown in Figure 4-18. The top trace shows the time evolution of the mean toroidal flux, and the bottom trace show the $m = 1$ magnetic field fluctuations. Prior to the crash, $m = 1$ precursor oscillations are observed in the soft x-ray flux and sometimes in the $m = 1$ edge magnetic probe. The toroidal flux generation indicates that plasma relaxation is occurring during the sawtooth crash phase.

In some experiments the sawtooth oscillations are smaller at lower values of Θ, the $m = 0$ magnetic perturbation becomes imperceptible, and continuous $m = 1$ oscillations appear. These characteristic oscillations are associated with dynamo activity, as discussed previously.

Theoretical Interpretation of the Sawtooth Crash

In a cylindrical plasma equilibrium, the mean magnetic field lines have a pitch that varies only with radius, as given by the function $q(r)$. Thus, in general, an equilibrium field line defines a surface of constant radius as it traverses the discharge. These surfaces are called flux surfaces, since they can be labeled by the amount of magnetic flux contained between the surface and the magnetic axis ($r = 0$). Rapid thermal conduction parallel to the magnetic field renders the temperature constant on a flux surface. The temperature will, of course, differ from surface to surface, with the inner surfaces being hotter. Magnetic confinement of a plasma can result from the relatively slow thermal transport across these surfaces.

Plasma relaxation occurs during the crash phase of a sawtooth oscillation. The RFP dynamo is operative during this phase. The dynamical processes that occur during such an event were described in Section 5.3. The resistive MHD modes that participate in the dynamo can cause magnetic islands to form at flux surfaces where they are resonant, i.e., where $q(r) = -m/n$. Modes with differing values of m/n are resonant at different radii. If the magnetic islands at neighboring resonant surfaces grow large enough to overlap with each other, the magnetic field lines in the region of overlap can become *stochastic*; they now fill a finite volume ergodically instead of being confined to a surface. The rapid thermal transport parallel to the field can now equilibrate the temperature throughout this volume,

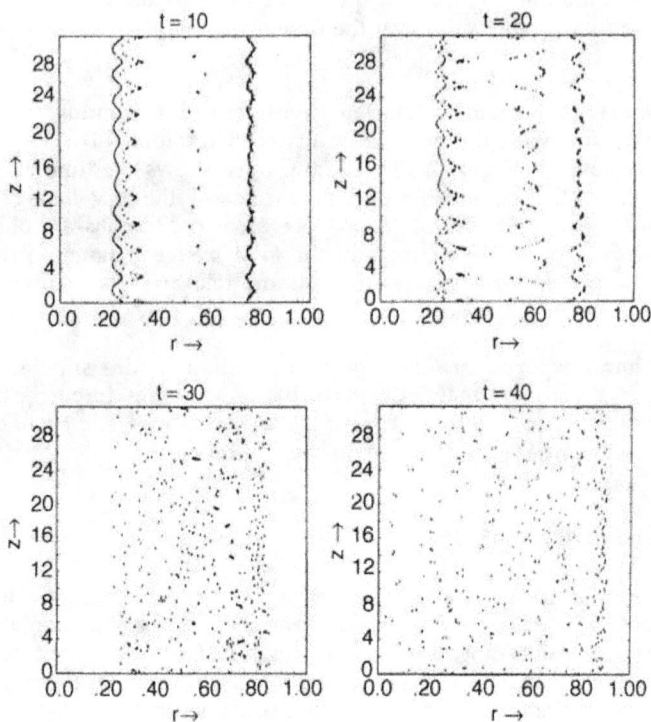

Figure 7-1. Generation of a region of stochastic magnetic field lines during a relaxation event at $\Theta = 1.5$. A mark is registered each time a field line pierces a given (r,z) surface. Note that the stochastic region is limited to $r/a < 0.9$, indicating possible edge confinement.

leading to a degradation of confinement [*Rechester and Rosenbluth*, 1978]. In some cases, the entire discharge may become stochastic.

It has been shown by means of numerical simulation [*Schnack et al.*, 1985] that the $m = 1$ activity responsible for the RFP dynamo leads to the destruction of flux surfaces and the appearance of stochastic magnetic fields in the core of the plasma. This is illustrated in Figure 7-1 for a case with $\Theta = 1.5$, and $R/a = 5$. These $m = 1$ modes couple nonlinearly with $m = 0$ modes that are resonant at the field reversal surface. At large Θ (or in the presence of a resistive wall), these $m = 0$ modes grow large enough to overlap with the $m = 1$ modes active in the plasma core. The discharge then rapidly becomes completely stochastic, as shown in Figure 7-2 for $\Theta = 1.8$. Such a state is

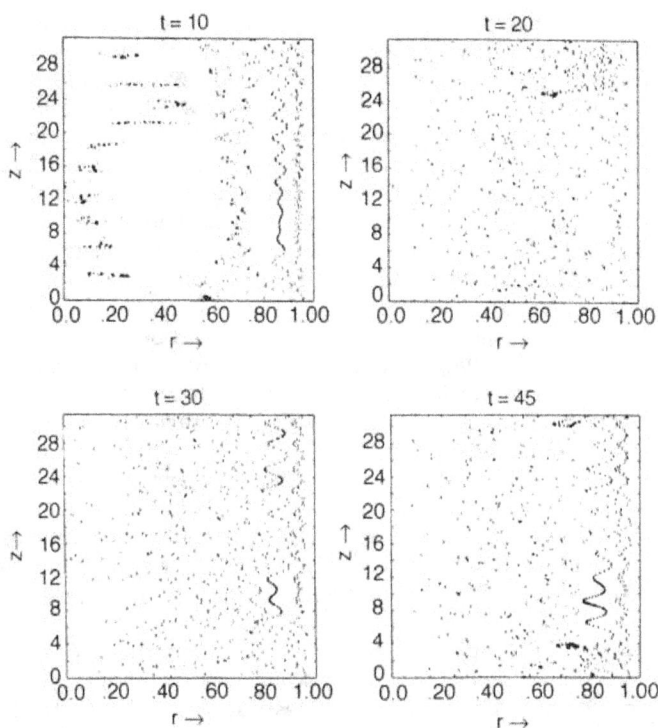

Figure 7-2. Generation of a region of stochastic magnetic field lines during a relaxation event at $\Theta = 1.8$. A mark is registered each time a field line pierces a given (r,z) surface. At this relatively large value of Θ the discharge can become completely stochastic $(t = 20)$. Note that good flux surfaces reappear at the edge later in time. (Compare with Figure 7-1.)

favorable for rapid outward transport of thermal energy. At low Θ (see Figure 7-1) the stochastic region is limited to the core, and good flux surfaces, and hence confinement, may remain at the plasma edge.

This picture of edge confinement is consistent with the observed sensitivity of the RFP to externally produced field errors. Toroidal flux generation, and hence dynamo activity, is experimentally observed to be associated with the sawtooth crash [Wurden, 1984]; similar flux generation has been observed numerically at high Θ [Schnack et al., 1985], and both experimentally and numerically when the current is increased at constant Θ [Phillips et al., 1988; Caramana and Schnack, 1986].

Sawtooth oscillations represent an extreme operating condition of the RFP dynamo, one in which the relevant time scales are sufficiently separated that their individual operation can be observed. Presumably, prior to the sawtooth ramp phase, the fluctuation level is low and the confinement is relatively good. Stochastic field lines are limited to some region in the core of the plasma, and the edge may provide good flux surfaces for confinement. Ohmic heating causes temperature and current density to gradually peak on-axis, thus lowering $q(0)$. When the q profile is sufficiently flattened, several $m = 1$ modes that cause plasma relaxation are destabilized. When these modes reach sufficient amplitude their nonlinear interaction causes the stochastic core to expand. Simultaneously, $m = 0$ magnetic islands are driven at the field reversal surface by the nonlinear interaction of the $m = 1$ modes. When these islands overlap with the $m = 1$ modes in the core the entire discharge may become stochastic causing rapid heat transport to the wall, lowering the core temperature, flattening the current, raising $q(0)$, and stabilizing the $m = 1$ modes. Simultaneously, the dynamo action driven by the modes causes a generation of toroidal flux and a restoration of the preferred, relaxed magnetic field profiles. After relaxation, the lowering of the $m = 1$ amplitudes causes the driven $m = 0$ modes to diminish; thus the overlap disappears, flux surfaces reappear at the plasma edge, and confinement improves. The process then repeats.

At low Θ, the amplitude of the MHD fluctuations responsible for plasma relaxation is lower, and the plasma edge may retain its integrity [*Schnack et al.*, 1985]. Also, the time scales governing the destabilization of the profiles by Ohmic heating, and the restoration of the profiles by plasma relaxation may not be widely separated in this parameter regime. The result is a quasi-continuous discharge that may not exhibit sawtooth oscillations.

The calculations reported in this section do not include thermal conduction. Hence, they are only suggestive of what might occur during the sawtooth crash. This deficiency is eliminated in the next section, where we describe similar calculations with self-consistent, classical Ohmic heating and anisotropic thermal conduction.

7.2 Thermal Transport During Sawtooth Oscillations

Energy Confinement Time

The scenario presented in Section 7.1 suggests that significant transport of thermal energy from the center to the edge of the plasma, where it is lost, may occur during the sawtooth crash that is associated with plasma relaxation. If the plasma is to remain hot, this lost energy must be resupplied by the external circuit in the form of electromagnetic energy. In the RFP, this is converted into thermal energy by means of Ohmic heating. The rate of transport of thermal energy from the core to the edge thus largely determines the heating power requirements of a particular experimental device. If this rate is too large, considerable input power may be required to maintain a specified plasma energy density.

The energy confinement properties of a plasma may be quantified by equating the power input to the plasma to the power loss due to thermal transport processes. Here, we assume that the only input power source is Ohmic heating. In steady state we thus write

$$\frac{\langle p \rangle}{\tau_E} = \langle \eta J^2 \rangle \qquad (7.1)$$

where $\langle .. \rangle$ represents a volume averaged quantity. The left hand side of Eq. (7.1) represents the power loss due to unspecified plasma transport processes. The quantity τ_E is the called *energy confinement time*, and can be used to quantitatively compare different plasma confinement systems.

Modifications to the Resistive MHD Model

Three-dimensional numerical simulations of sawteeth with finite plasma pressure have been performed with self consistent models for resistivity and anisotropic thermal conductivity. Pressure forces are included in the equation of motion and the energy equation

$$\frac{\partial p}{\partial t} = -\gamma p \nabla \cdot \mathbf{v} - \mathbf{v} \cdot \nabla p + \frac{2(\gamma - 1)}{S \beta_0} \eta J^2 - (\gamma - 1) \nabla \cdot \mathbf{q} \qquad (7.2)$$

is solved. The first term on the right hand side of Eq. (7.2) describes adiabatic heating or cooling due to compression or dilation, the second term gives the convection of thermal energy due to plasma flow, the third term gives the increase in thermal energy due to Ohmic (resistive) heating, and the fourth term represents the effects of thermal conduction. The normalized resistivity is assumed classical, so that $\eta(T) = T^{-3/2}$. The heat flux vector $\mathbf{q}$ is given by

$$\mathbf{q} = -\kappa \cdot \nabla T , \qquad (7.3)$$

where the thermal conduction tensor κ is expressed in a coordinate system locally aligned with the magnetic field $\mathbf{B}$ as

$$\kappa = \begin{pmatrix} \kappa_\perp & 0 & 0 \\ 0 & \kappa_\perp & 0 \\ 0 & 0 & \kappa_\parallel \end{pmatrix}, \tag{7.4}$$

where $\kappa_\perp$ and $\kappa_\parallel$ are given in nondimensional form by

$$\kappa_\perp = 4.97 \, \beta_0 \, \frac{n_0^2}{B^2 T^{1/2}}, \tag{7.5}$$

and

$$\kappa_\parallel = 4.97 \, \beta_0 \, C_\parallel \, T^{5/2}. \tag{7.6}$$

The constant $C_\parallel$ is related to the normalization values, and is given by

$$C_\parallel = 2.36 \times 10^{22} \, \frac{B_0^2 T_0^3}{n_0^2}. \tag{7.7}$$

It is obvious from the temperature dependence of Eq. (7.6) and Eq. (7.7) that, in a high temperature plasma, parallel thermal conduction is much more rapid than perpendicular conduction. However, in a plasma with nested flux surfaces, the *radial* transport of energy will be due to perpendicular conduction alone; parallel conduction serves merely to equilibrate the temperature along the flux surfaces. The radial transport thus predicted is generally too small to account for experimental observations.

Equation (7.3) can be expressed in the laboratory frame by a simple similarity transformation which is dependent upon the local magnetic field vector **B**.

In this model the density remains constant, so the plasma pressure can be identified with the plasma temperature.

Simulation of Sawtooth Oscillations

Using this model the start-up, formation, and sustainment of a cylindrical RFP discharge from a cold, zero-current, uniform flux state has been studied. The initial state is meant to represent a cold, ionized, uniform plasma with an imbedded uniform toroidal (axial) magnetic field. This plasma initially carries no current. The spatially uniform initial conditions were $J_\theta = J_z = 0$, $B_z = 1$ KG, $n_0 = 10^{14}$ cm^{-3}, and $T_0 = 13.6$ eV. The minor radius was $a = 20$ cm and the aspect ratio was unity. These conditions result in $S = 2.6 \times 10^3$, $\beta_0 = .055$, and $C_{\parallel} = 5.9 \times 10^3$. The toroidal flux was held constant throughout the calculation while the toroidal current was linearly raised from zero to a value corresponding to $\Theta = 1.8$, and then held fixed in time. This calculation thus models the start-up, self reversal, and sustainment phases of an RFP discharge.

The time evolution of the discharge is characterized in Figure 7-3, where we plot the field reversal parameter F as a function of normalized time. (Time is normalized to the resistive diffusion time in the initial, cold, current-free state τ_{R0}.) In this figure, the pinch parameter Θ reaches its maximum value of 1.8 at a normalized time of 0.35. The initial current rise causes a compression that leads to a diamagnetic but non-reversed pinch. This state is unstable to $m = 1$ resistive MHD modes. At approximately $t/\tau_{R0} = 0.8$ these modes have grown sufficiently large to cause field reversal.

The final sustained state is characterized by repeated sawtooth oscillations, with an average value of the poloidal beta parameter $\beta_p = 8\pi \langle p \rangle / B_\theta(a)^2 \approx 0.1$, a Lundquist number $S = 1.7 \times 10^5$, a toroidal current of 180 KA, and an average energy confinement time, Eq. (7.2), of $\tau_E \approx 1.8$ msec.

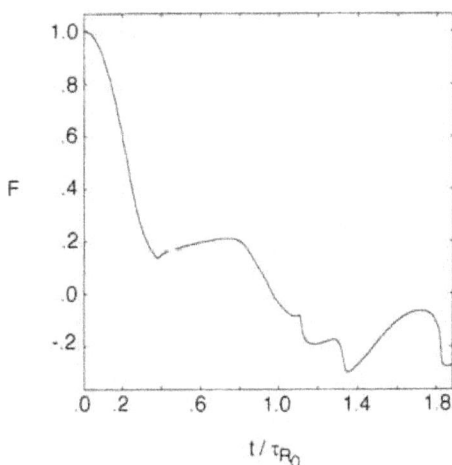

Figure 7-3. Field reversal parameter F vs. time (measured in units of the resistive diffusion time in the initial state) from a three dimensional numerical simulation of RFP start-up, self-reversal, and sustainment. The model contains finite plasma pressure. Reversal occurs at $t/\tau_{R0} \approx 1$. Note the sawtooth oscillation that occurs between $t/\tau_{R0} \approx 1.4$ and $t/\tau_{R0} \approx 1.8$.

These values are in at least qualitative agreement with experimental results [*Massey et al.*, 1985; *Hokin et al.*, 1991].

In Figure 7-4 we plot the $m = 0$, $n = 0$ axial magnetic field at the conducting wall $r = a$ as a function of time for one cycle of the sawtooth oscillations that occur during the sustainment phase of the simulation. Since the toroidal flux was held constant throughout the calculation, plasma relaxation and dynamo action during the sawtooth crash phase is characterized by enhanced field reversal, as shown in Figure 7-4. Figure 7-5 shows the corresponding mean pressure (or temperature, since the density was not evolved) at the magnetic axis $r = 0$. Note the asymmetry of the oscillations, exhibiting a clear separation between the heating and crash phases.

During the slow rise phase of the sawtooth, Ohmic heating dominates and the dynamo fluctuations decrease. This results in an increase in the central pressure (implying improved confinement), and a decrease in field reversal (the axial field at the wall becomes less negative). As the mean fields are modified, the current driven kink modes responsible for the dynamo are destabilized, the fluctuation level increases, and the central pressure rapidly decreases (implying degraded confinement). The mean pressure profile immediately after the sawtooth crash is shown in Figure 7-6; it is flat over nearly ninety percent of the discharge. The magnetic field lines exhibit a stochastic variation similar to that shown in Figure 7-2.

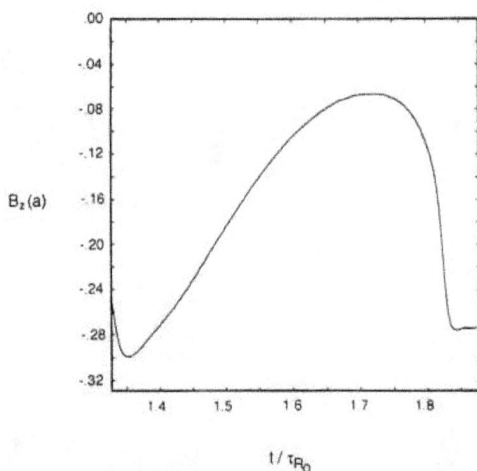

Figure 7-4. Average axial magnetic field at the wall (measured in units of the initial axial magnetic field on axis) vs. time (measured in units of the resistive diffusion time in the initial state) for a sawtooth oscillation during the sustainment phase.

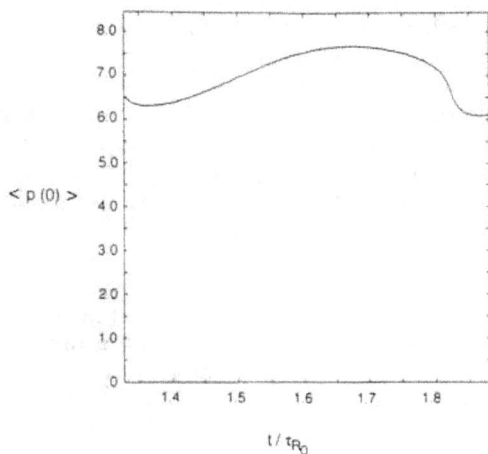

Figure 7-5. Average plasma pressure on axis (measured in units of the initial pressure on axis) vs. time (measured in units of the resistive diffusion time in the initial state) for a sawtooth oscillation during the sustainment phase.

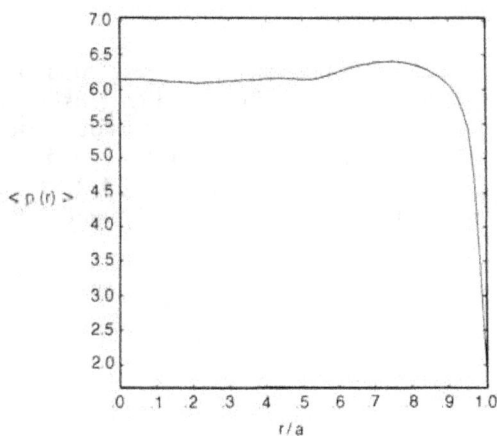

Figure 7-6. Mean plasma pressure (measured in units of the initial pressure on axis) vs. radius immediately after a sawtooth crash.

7.3 Summary

The MHD fluctuations that cause plasma relaxation, and hence sustainment of the discharge, can also cause regions of stochastic magnetic field lines. Classical thermal conduction along these field lines can cause the plasma temperature to equilibrate throughout a three-dimensional volume of the discharge. The result is that the net transport of thermal energy from one macroscopic region of the plasma to another may be faster than might be expected on the basis of classical cross-field transport, and is dominated by parallel transport along the stochastic field lines.

Reversed-field Pinch experiments exhibit sawtooth oscillations in both soft x-ray emission and edge toroidal magnetic field in at least some operating regime. These oscillations have a slow rise phase and a rapid crash phase. During the rise phase, the magnetic fluctuations are low and the soft x-ray emission increases due to Ohmic heating. The central current density also increases because of the inverse dependence of the resistivity on temperature. The resulting modification of the magnetic field profiles drives the discharge away from its relaxed state and destabilizes global resistive MHD modes. When these modes reach sufficient amplitude their nonlinear interaction causes rapid plasma relaxation and a restoration of the preferred, relaxed magnetic field profiles; the discharge is sustained. Concurrently, the resulting stochastic magnetic field lines cause a rapid transport of thermal energy from the core of the plasma to the cold edge, where it is lost. This is the crash phase of the sawtooth oscillation.

It is important to note that the calculation presented in Section 7.2 contains only classical transport coefficients [*Braginskii*, 1965]. The anomalously rapid radial heat transport is due to parallel heat conduction along the self-consistently calculated stochastic magnetic field lines, as given by the anisotropic heat conduction tensor, Equations (7.4–7.6). Anomalously large local transport coefficients (i.e., not based on Coulomb collisions) may not be required to explain thermal transport.

CHAPTER 8

DYNAMICAL RELAXATION IN THE SOLAR CORONA

In the previous chapters we have presented both a general theory and specific calculations related to plasma relaxation. The calculations presented in Chapters 5 through 7 pertain to the specific example of the Reversed-field Pinch laboratory experiment. Because this magnetoplasma configuration has been studied in detail, we used it as a paradigm for understanding the dynamics of plasma relaxation.

We now briefly turn our attention to a naturally occurring magnetized plasma: the solar corona, or outer atmosphere of the sun. Because of its proximity to earth, this plasma has been extensively observed. The coronal magnetic field is anchored in the photosphere, or dense visible surface of the sun. This forms the lower boundary of the coronal plasma.

Like the RFP, the corona has been found to be highly dynamic. It exhibits both quasi-continuous and large amplitude quasi-periodic magnetic fluctuations. Also like the RFP, it is driven away from its preferred state by electric fields at its boundary. For the case of the RFP, this electric field is imposed by the external circuit. For the case of the solar corona, this field is induced by convective motions in the photosphere that move the feet of the coronal magnetic field lines about. In both cases, the resulting dynamics can be interpreted as a form of plasma relaxation. The dynamics of the solar corona are likely similar to those occurring in stellar coronae throughout the universe.

In this chapter we present the results of numerical simulations, similar to those presented in Chapter 5, that are relevant to the dynamics of the solar corona. We begin with an overview of coronal dynamics and then present the details of two numerical simulations of the corona. One is related to the dynamics of active regions, and is reminiscent of the relaxation processes that occur during the crash phase of a sawtooth oscillation in the RFP. The second is related to the dynamics of the quiet corona, and is relevant to theories of coronal heating.

8.1 Overview of Coronal Dynamics

The solar corona, or outer atmosphere of the sun, consists of a hot, tenuous plasma permeated with a magnetic field. This magnetized plasma displays a rich variety of dynamical activity. The coronal magnetic field is rigidly attached to the photosphere, the visible solar surface. The footpoints of these field lines are constantly being moved about slowly by the motion of the photosphere, which is driven from below by convection. This motion produces a continual stirring of the coronal plasma. New magnetic flux emerges from below the photosphere and displaces plasma above it. Sunspots and their associated strong magnetic fields interact with each other, causing the magnetic field lines to slowly twist into loops and arcades resembling croquet hoops. Immense filaments of dense, cool plasma mysteriously form within the corona, and persist for periods of days or weeks. Occasionally these slow, churning processes yield to rapid, dramatic events. The local magnetic structure of the corona may be suddenly altered, with the release of vast amounts of energy in a very short period of time. A broad band of radiative energy is released in the form of a solar flare. A cool, dense filament may erupt from the low corona. Rapid ejections of mass into interstellar space may accompany topological changes in the coronal magnetic field. These events may occur in concert or individually, and often seem to emerge spontaneously and without warning from the relatively quiescent corona.

A quantitative theoretical description of the dynamical corona is one of the main challenges of modern solar physics. It is generally believed that the source of the energy released in disruptive events such as solar flares, erupting prominences, and coronal mass ejections is the coronal magnetic field. The energy in this field has two components. One is due to electric current flowing below the photosphere. This energy generally cannot be tapped as this would require the alteration of distant current sources. The other arises from electric currents flowing locally in the corona itself. This energy *can* be tapped by the local rearrangement of magnetic field. The origin of this free energy is likely the work done on the field by the slow motion of the photospheric surface. The central question is how large amounts of magnetic energy can be self-consistently stored in the corona over relatively long periods of time, and then rapidly released.

An analogy can be drawn between the dynamics of the Reversed-field Pinch and the solar corona. Both are externally driven magnetoplasma systems that evolve away from their preferred states. In both cases, the external driving agency is an applied electric field. In the case of the RFP, this is the toroidal voltage supplied by the external circuitry. The RFP plasma evolves away from its preferred, relaxed state as a result of resistive diffusion; the applied voltage is just sufficient to provide the free energy necessary to drive the dynamical relaxation process that has been described in detail in Chapter 5. In the case of the solar corona, the applied electric field is induced by convective motions in the bounding photospheric surface that is penetrated by the coronal magnetic field. Since resistive diffusion is exceedingly slow in the coronal plasma, this applied electric field must drive the coronal field away from its relaxed state in addition to supplying the free energy required for dynamical relaxation.

The corona is observed to consist of relatively small regions of intense magnetic activity imbedded in a quiescent, hot plasma. These small, intense regions are called *active regions,* and are characterized by complex magnetic field geometry and rapidly evolving structures. These regions are often associated with sunspots, and are the site of solar flares and erupting prominences. The photospheric motions in these regions may exhibit considerable variation (shear) on long length scales. In these regions the coronal magnetic fields may be strongly driven by these coherent motions. In contrast, the quiet, ambient corona is characterized by relatively simple large scale magnetic fields, and by photospheric flows that exhibit little long range structure. In these regions, which constitute the majority of the corona, the feet of the field lines are merely shuffled in a random manner.

In the following sections we will briefly describe detailed calculations relevant to the dynamics of both active and quiescent regions of the corona. These calculations use the same techniques of numerical simulation that have been applied to the RFP. For the active region we present a model problem that generically models the dynamics of a magnetic arcade, a structure that is characteristic of many active regions. This problem demonstrates that magnetic energy can be slowly stored in the corona and then rapidly released. This rapid release of energy is a relaxation event, analogous to a sawtooth oscillation, and may be the source of energy for solar flares. For the quiet corona, we present a model problem that demonstrates the response of a simple coronal field to random photospheric motions. These results are relevant to models of coronal heating, and may represent a form of continuous plasma relaxation.

We remark that it is difficult in these cases to make a direct connection with Taylor's relaxation theory; the situation is complicated by the boundary conditions at the photosphere and at infinity. Nonetheless, the plasma dynamics are similar to those observed in the RFP.

8.2 Magnetic Arcade Evolution

Among the features of active solar regions is the presence in the photosphere of adjacent areas of oppositely directed component of the magnetic field normal to the photosphere. (This field component is often referred to as the line-of-sight field, since it is only the component along the line-of-sight from earth that can be measured.) These regions are separated by a neutral line, where the normal component of the field vanishes. In the absence of coronal electric currents, the corresponding potential (i.e., current-free) coronal magnetic field forms an arcade spanning the neutral line. The projection of this field in the plane of the photosphere (the transverse field) is normal to the neutral line. If motions occur in the photosphere that are sheared with respect to the neutral line (as is generally true in developing active regions), the ensuing motion of the footprints will stretch the magnetic arcade along the neutral line. They will also induce electric currents in the corona, increasing the magnetic energy above the value of the potential (current-free) field. The transverse field will then make an acute angle with respect to the neutral line.

Observations have indicated a strong correlation between the deviation of the angle between the transverse field and the neutral line from its potential field value and the onset of solar flare activity. Using detailed observations of fibril changes in the active region McMath 14943 (September 1977), Neidig, DeMastus, and Wiborg [Neidig et al., 1978] deduced energy build-up in sheared (non-potential) fields prior to both small and large flares, and identified a relaxation in shear following these flares. They further concluded that flares were associated with relatively large regions of emerging flux and shear, rather than with small scale events. Direct evidence of the correlation between stressed magnetic fields and flares has been given by Hagyard, et al. [Hagyard et al., 1984] using a vector magnetograph, which measures the transverse photospheric magnetic field. In studies of active region AR2372 (April 1980) they observed regions of both potential and sheared field. Flares occurred only in the region of shear. Additionally, they observed that the shear persisted during several flare events, and suggested that there exists a critical value of shear (characterized by the angle θ between the transverse field and the neutral line) that triggers the flare event ($\theta < 5 - 10°$). They envisioned that photospheric flow deforms the field until shear exceeds a critical value, resulting in a flare that causes the relaxation of the shear to a somewhat smaller value. Additional deformation can induce critical shear to be exceeded again, explaining the observed tendency for repeated flaring in active regions.

The response of a model arcade field to shearing footpoint motion can be studied solving the force free model of Chapter 2 as an initial value problem, with the proper boundary conditions imposed at the photosphere [Mikić et al., 1988]. The model configuration chosen for the initial condition is shown in Figure 8-1. In this Cartesian coordinate system, x is the altitude above the photosphere, y is the direction transverse to the neutral line, and z is the direction parallel to the neutral line. This initial potential field is periodic in the transverse (y) direction, and is given by

$$B_x = -B_0 e^{-k_1 x} \sin(k_1 y) , \tag{8.1}$$

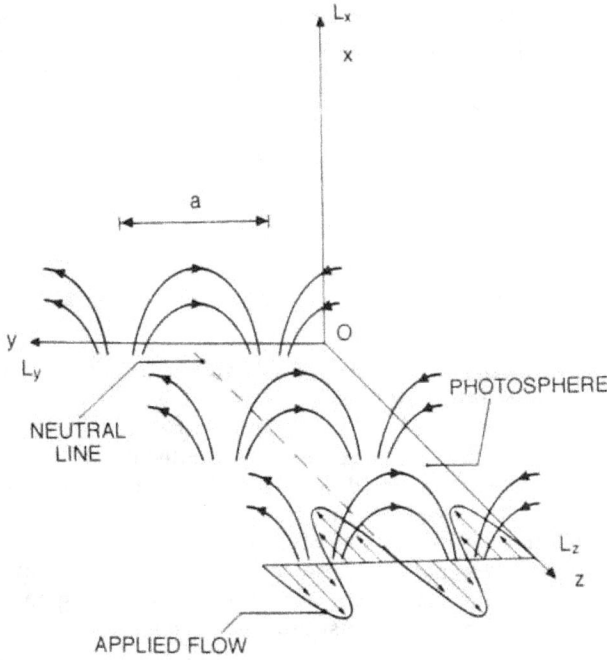

Figure 8-1. Sketch of magnetic arcade geometry. The system is periodic in both the y and z coordinates.

$$B_y = B_0 e^{-k_1 x} \cos(k_1 y) , \qquad (8.2)$$

$$B_z = 0 , \qquad (8.3)$$

for $0 \leq y \leq L_y$, $0 \leq z \leq L_z$, where $k_1 = 2\pi/L_y$. Periodicity in y implies that this condition represents an infinite array of arcades arranged side by side and extending in the z-direction.

The effect of the photospheric flow is modeled by specifying the y and z components of the velocity at $x = 0$ as functions of time. A particularly simple model is two dimensional, with $v_y = 0$, and v_z given by

$$v_z = -v_0 f(t) \sin(2k_1 y) , \qquad (8.4)$$

where $f(t)$ is a ramp function that is increased linearly to unity in time t_R, and then fixed. The ramp time t_R is taken to be longer than the Alfvén time to reduce transients. The line-tied boundary condition at the photosphere is enforced by holding the normal component of the magnetic field fixed in time, and computing the tangential electric field from Ohm's law. The resulting boundary conditions are

$$B_x(0,y,t) = B_x(0,y,t = 0) , \tag{8.5}$$

$$E_z(0,y,t) = 0 , \tag{8.6}$$

$$E_y(0,y,t) = -v_z(0,y,t) \, B_x(0,y,t) , \tag{8.7}$$

The goal of model simulations of this sort is to consider the simplest dynamical model that describes the generic behavior of representative magnetic configurations. Since the magnetic forces dominate the pressure forces in the corona, the pressure force is neglected in the momentum equation. Furthermore, since the dynamics of interest is associated with the magnetic evolution, it is assumed that the mass density is uniform in space and constant in time. The further neglect of gravitational effects is thus appropriate. Viscous dissipation is also added to the momentum equation to control short wavelength oscillations and allow for damping to steady state solutions. Even with these assumptions, rich dynamical behavior reminiscent of solar observations occurs.

The physical parameters for the arcade evolution are given in Table 8-1. With these parameters, we find the magnetic energy in the initial potential field, Eqs. (8.1–8.3), to be $W_{pot} = 4 \times 10^{31}$ ergs. We also note that the photospheric flow velocity of 31 km/sec is an order of magnitude or more larger than the solar value. It has been found that the results described below are independent of the actual value of the photospheric velocity, provided that this value is significantly less than the Alfvén velocity; smaller values merely make the computations more costly.

Table 8-1. Summary of Physical Parameters for the Arcade Simulation

Parameter	Value
Arcade Width, a	50,000 km
Magnetic field at base, B_0	100 G
Plasma density, n_e	5×10^9 cm^{-3}
Alfvén time, τ_A	32 sec
Resistive diffusion time, τ_R	3.2×10^5 sec
Viscous diffusion time, τ_v	3.2×10^3 sec
Lundquist number, S	10^4
Alfvén speed at base, v_A	3087 km sec^{-1}
Maximum photospheric flow velocity, v_0	31 km sec^{-1}

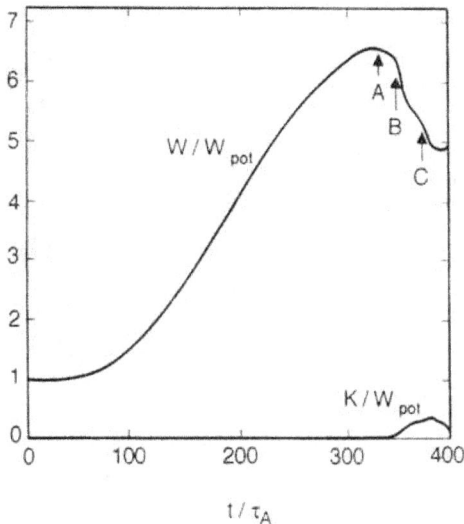

Figure 8-2. Magnetic energy and kinetic energy (normalized to the magnetic energy in the initial potential magnetic field) vs. time (measure in units of Alfvén time) during the evolution of a sheared magnetic arcade.

The evolution of the total magnetic and kinetic energies as a function of time is shown in Figure 8-2. It is found that the evolution of the arcade proceeds through two stages. The first is a quasi-static phase in which the energy slowly builds to a value over six times the initial potential energy. This is followed by a rapidly evolving dynamic phase in which a significant amount of magnetic energy is converted into kinetic energy and heat. For the calculation described here, the energy buildup occurred over a period of about three hours; with a realistic value of the photospheric flow velocity, this time would be 10 to 100 times longer. During the rapid dynamical phase, approximately 4×10^{31} ergs is released from the magnetic field in about 15 minutes.

Both top and end-on projections of magnetic field lines during the arcade evolution are shown in Figure 8-3. The quasi-static phase is characterized by a slow rising and stretching of the arcade magnetic field lines as the footpoints are sheared. The rapid dynamical phase is characterized by a twisting deformation, and the ejection of a plasmoid from the corona. A detailed examination of this phase shows that a current sheet forms in the middle of the central arcade. The plasmoid is ejected as a result of resistive magnetic reconnection across this interface. The onset of the dynamical phase is found to correspond to the growth of a global ideal MHD instability

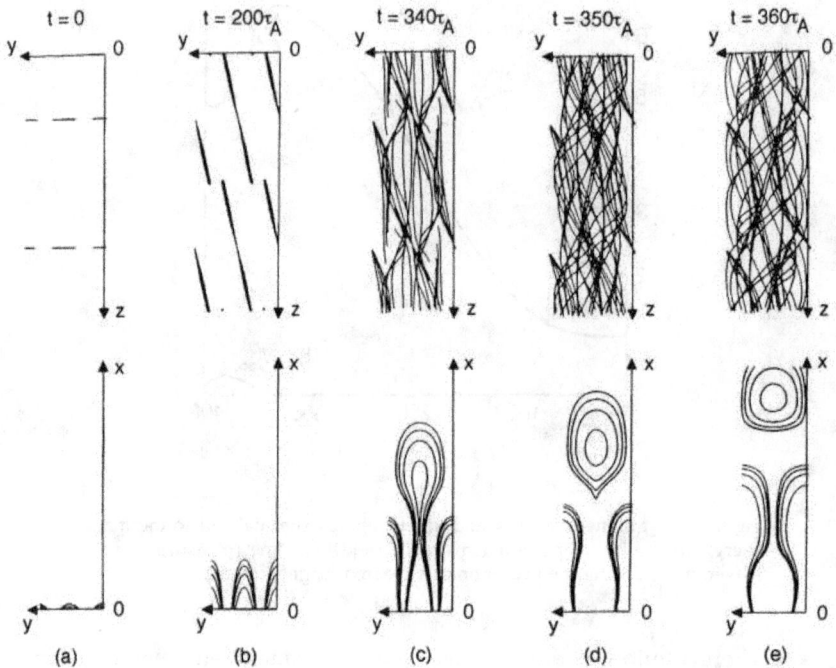

Figure 8-3. Evolution of the arcade magnetic field lines. The times correspond to times shown in Figure 8-2. A plasmoid is ejected late in time (c-e).

of the arcade configuration. For the present geometry, this mode becomes unstable when the magnetic energy exceeds about four times the energy in the initial potential field.

The results described above have several features in common with previous solar flare models [*Sturrock*, 1980], and with observations. These results demonstrate that a slow buildup of magnetic energy can occur in the corona as a result of electric currents induced by photospheric motions. This energy can be stored in the corona over a long period of time (days to weeks). If a critical amount of shear is exceeded, the configuration becomes linearly unstable on the Alfvén time scale. The nonlinear evolution of the instability causes current sheets to form. Reconnection of the magnetic field produces a rapid release of magnetic energy roughly equal in magnitude to the energy in a potential field with the same normal component of the magnetic field in the photosphere. Reconnection in the central arcade produces an opening of the closed field lines and ejects a plasmoid upward at the local Alfvén velocity. Some of the magnetic energy is converted into kinetic energy of the plasmoid; the remainder is dissipated as Joule and viscous heating near the current sheet.

Similar simulations have been performed [Biskamp and Welter, 1989], but without the periodic transverse symmetry reported above. It was found that the shearing of only one or two isolated arcades does not lead to disruption, but rather to a continual expansion of the arcade fields into the corona. Only when three neighboring arcades were sheared did the central arcade disrupt in a manner similar to that reported in this section. This is reminiscent of solar observations, where disruptive behavior is usually associated with complex flux systems. Perhaps the interaction with neighboring flux systems is a necessary ingredient for the onset of disruptive coronal dynamics.

8.3 Coronal Current Filaments

In Section 8.2 possible self-consistent mechanisms for energy storage and conversion in the solar corona were presented. The problem of the quiet corona, in particular the response of the coronal field to *random* footpoint displacements produced by photospheric motion, is now described. The results will have impact on models of steady state coronal heating.

The physical processes that heat the solar and other stellar coronae are not well understood [Priest, 1982]. However, it is widely accepted that the magnetic field in the solar atmosphere plays a prominent role in the energy deposition process. Based on this concept, it has been suggested that the corona is heated through the dissipation of electric currents. In one version of this picture, the shearing and compression of a large scale field whose footpoints are anchored below the surface can occur as a result of convective motions in the photosphere. This deformation induces electric currents to flow in the corona. The length scale on which the footpoints are shuffled by the convective motions is approximately $l_v \sim 10^6$ m. If the coronal heating rate is to be accounted for by Joule dissipation of electric currents, the magnetic field is required to have structure at scale lengths several orders of magnitude smaller than l_v, due to the large value of the conductivity in the high temperature corona. Thus, in order to explain coronal heating by the above mechanism, it is required to demonstrate how the magnetic field can acquire structure at length scales $l \ll l_v$.

The mechanism by which the magnetic field acquires small-scale structure is somewhat controversial. Parker [*Parker*, 1972, 1983, 1986] has noted from a perturbation analysis that, in general, footpoint displacements lead to a loss of equilibrium, inducing current sheets to form in the corona. In an ideal (perfectly conducting) plasma, these sheets are mathematical discontinuities of infinitesimal thickness.   In a more realistic resistive plasma, these sheets attain finite, but still small, thickness.   The rapid resistive reconnection of magnetic field at these sheets would provide enhanced dissipation over that produced by normal Joule dissipation of the large-scale field, and would allow the field to reach equilibrium with a much simpler topology. In Parker's hypothesis the structure at $l << l_v$ in the corona is formed trivially through the creation of these current sheets resulting from loss of equilibrium.

In contrast, other authors [*van Ballegooijen*, 1985, 1988a; *Antiochos*, 1987; *Zweibel and Li*, 1987] have noted that well-behaved, continuous photospheric flows produce coronal fields which are free of discontinuities in the absence of initial neutral points in the field. However, it has been hypothesized [*van Ballegooijen*, 1985] that random smooth flows (which are characteristic of the flows produced by subphotospheric convection) produce structure in the magnetic field at arbitrarily small length scales in a nonlinear cascade process, similar to that described in Chapters 1 and 2. A statistical analysis [*van Ballegooijen*, 1986] has determined the rate at which the structure at l_v is expected to cascade to short length scales. For an idealized random flow profile it was conjectured [*van Ballegooijen*, 1986] that the electric current density in the corona ought to build up exponentially in time.   Indeed, a realization of the first few steps of this process [*van Ballegooijen*, 1988a], involving the numerical solution of the *equilibrium* MHD equations, indicates that such flows rapidly introduce fine structure in the coronal field.

In either model, coronal heating results from the rapid dissipation of these small scale electric current structures by electrical resistivity.

In this section dynamical calculations that address the issue of the creation of small-scale structure in magnetic fields are described [*Mikić et al.*, 1989]. The time-dependent three-dimensional *ideal* (η = 0) MHD equations in a geometry that is an idealization [*Parker*, 1972] of the solar corona are solved. As before, the MHD equations are solved in the limit of vanishing gas pressure and constant mass density. Gravitational effects are ignored. It is found that a sequence of smooth, randomly phased flows generates structure in the field at increasingly shorter lengths. This structure is evidenced by an exponentially growing current density and a cascade of spatial structure to smaller scales. The rate of generation of the small scale electric current density is consistent with statistical predictions [*van Ballegooijen*, 1988a]. It is found that the field relaxes to equilibria that agree with those computed from equilibrium considerations [*van Ballegooijen*, 1988b]. The dynamical evolution shows that the field is able to reach equilibrium at each step of the sequence of applied motions *without* producing discontinuities in the field. True discontinuities would thus arise only after an infinite number of steps. However, current structure at arbitrarily small scale length can be generated in finite time.

The dynamical MHD equations are used as a convenient way of imposing a sequence of applied motions on the field footpoints. The real time evolution of the coronal field as it would respond to continually moving footpoints is not followed since the prime interest is in the equilibrium solutions of the equations. An enhanced value of the plasma viscosity is used to look for equilibrium solutions. This enables the build-up of structure in the equilibrium field as a result of each applied footpoint displacement to be studied. Nevertheless, the present treatment does provide information on the stability of resulting equilibria, since the inertial term in the momentum equation is retained.

It is assumed that the initial coronal field may be represented by a uniform field $\mathbf{B}_0 = B_0 \hat{x}$ extending between two infinitely conducting plates at $x = 0$ and $x = L_x$ which represent the two ends of a coronal loop anchored to the photosphere [*Parker*, 1972]. We use line-tied boundary conditions [*Van Hoven et al.*, 1981] at $x = 0$ and $x = L_x$ which prescribe the motion of the field line footpoints. For convenience, we assume that the field lines do not move at $x = L_x$, and we apply the desired flows at $x = 0$. The line tied boundary condition is implemented.

The force-free MHD equations, with the resistivity η set to zero, are solved in Cartesian coordinates in a cube of dimension $L_x \times L_y \times L_z$. The y and z directions are assumed to be periodic. These equations are solved numerically for a specified footpoint motion $\mathbf{v}(y,z,t)$ at $x = 0$. An enhanced value of the plasma viscosity is used to relax the field to equilibrium, if one exists, while allowing only ideal motions. This provides an accelerated (to save computer time) simulation of the viscous equilibration of slow, photospherically driven motions. At the upper boundary $x = L_x$, $\mathbf{v} = 0$. At the lower boundary $x = 0$, $v_x = 0$, and a sequence of tangential flows, for $i = 1, 2, 3,...$, is applied [*van Ballegooijen*, 1988a, 1988b]:

$$\left. \begin{aligned} v_y &= 0 \\[1em] v_z &= v_0 \, \sin\left(k_1 y + \phi_i\right) \end{aligned} \right\} \quad \text{for odd } i, \qquad (8.8)$$

and

$$v_y = v_0 \sin (k_1 z + \phi_i)$$

$$v_z = 0$$

for even i. (8.9)

This flow pattern contains spatial structure at wave number $k_1 = 2\pi/L$. The phases ϕ_i of the above flows are selected randomly (distributed uniformly between 0 and 2π). The magnitude of the flow v_0 is chosen so that $k_1 v_0 T = 1$, where T is the time during which the flow is applied. In the simulations T is chosen to keep v_0 a small fraction of v_A in order to minimize the introduction of transients. A sequence of 15 random phases are applied. After each step in the sequence, the field relaxes towards an equilibrium as a result of viscous dissipation, and the velocity is damped to $\sim 10^{-5} v_A$ before the next flow pattern is applied.

Figure 8-4 shows the traces of 16 representative field lines after the field has settled to equilibrium, at step $i = 0$ (corresponding to the initial field), and at steps $i = 4, 8$, and 12. The same set of field lines is followed in this sequence. It is apparent that the randomly phased flow causes the individual field-line footpoints to wander in a "random walk" fashion. Figure 8-5 shows the build-up of the current density, in the form of contours of μ in the y-z plane at $x = L/2$, at steps $i = 4, 8$, and 12. The normalized parallel current density μ (defined by $J = \mu B$) shows increasing transverse structure as i increases. Since the large-scale field remains primarily in the vertical direction, μ is indicative of the vertical current density, J_x. The vertical structure of μ is shown in Figure 8-6. This figure shows a three-dimensional projection of a surface of constant μ at step $i = 12$. Taken together, Figures 8-5 and 8-6 show that the variation of μ is primarily in the transverse direction. It is evident that μ (and hence the electric current density) develops an increasingly finer spatial structure in the transverse plane as i increases.

If this mechanism were operative in the solar corona, the magnetic field would have a fine-scale structure superimposed on the large-scale field, and the electric current density would have a filamentary structure. The continuous motion of the magnetic field footpoints due to subphotospheric convection would distort the coronal field. The structure introduced on the convective length scale l_v would cascade to shorter length scales, as seen in this section, thereby creating current filaments which would be dissipated by resistive diffusion once they reached a sufficiently small spatial scale. This continuous churning of the field could provide a statistically steady heating source.

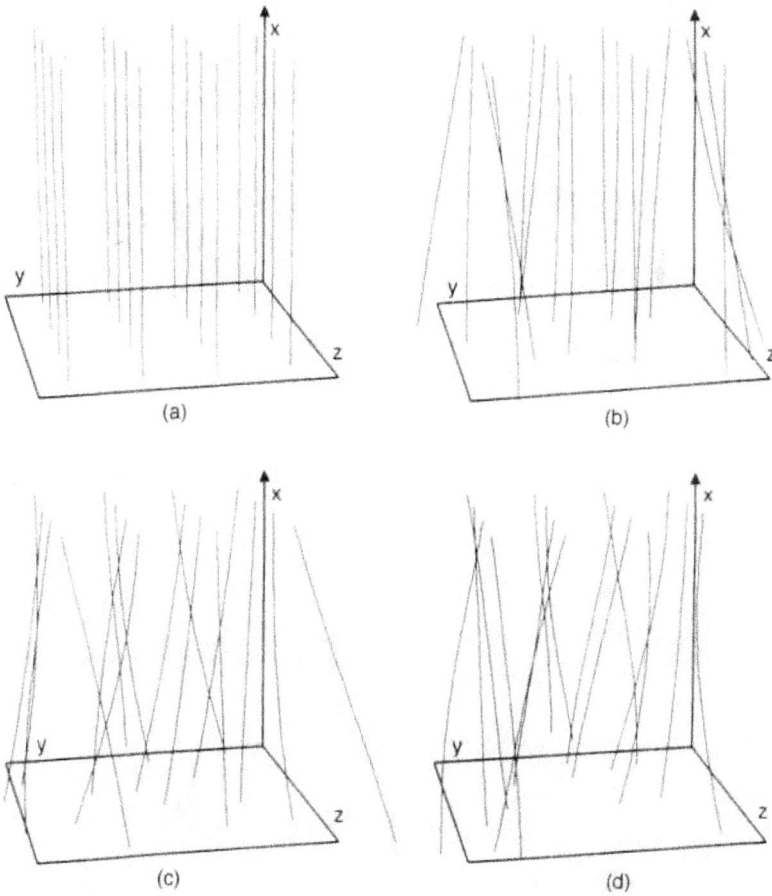

Figure 8-4. Magnetic field lines produced by random shuffling of the feet in the lower plane.

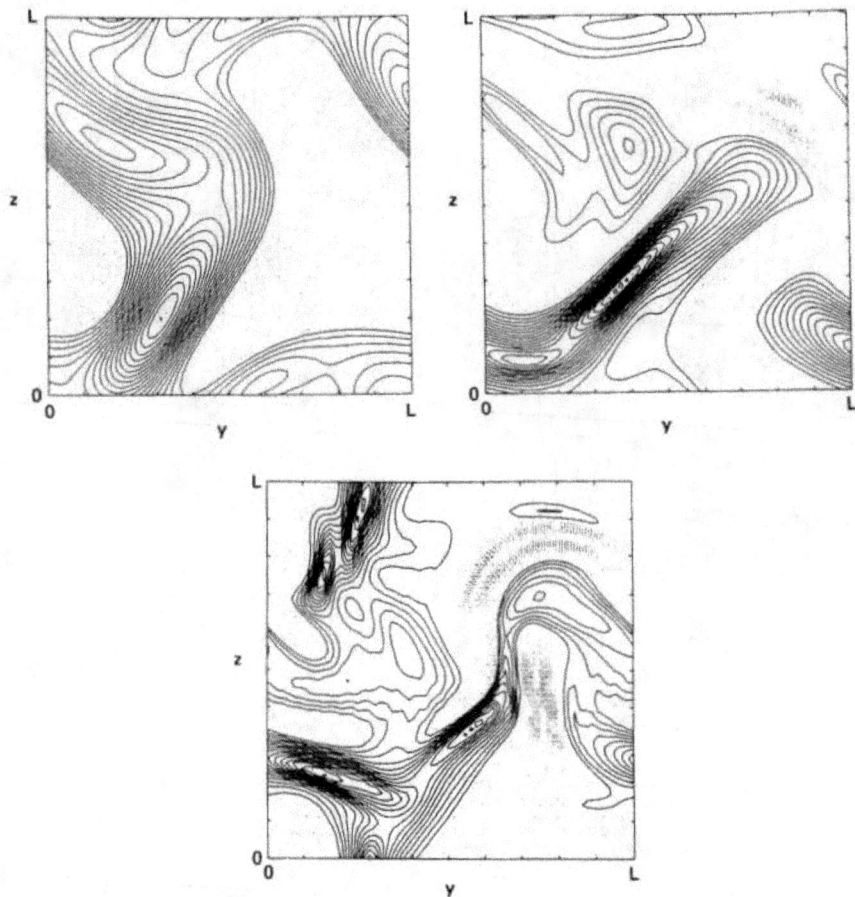

Figure 8-5. Contours of normalized parallel current density in the (y,z) plane at $x = L/2$ during the random shuffling process. Note that current structures are formed that have much smaller spatial scales than the flows that moved the foot points of the field lines.

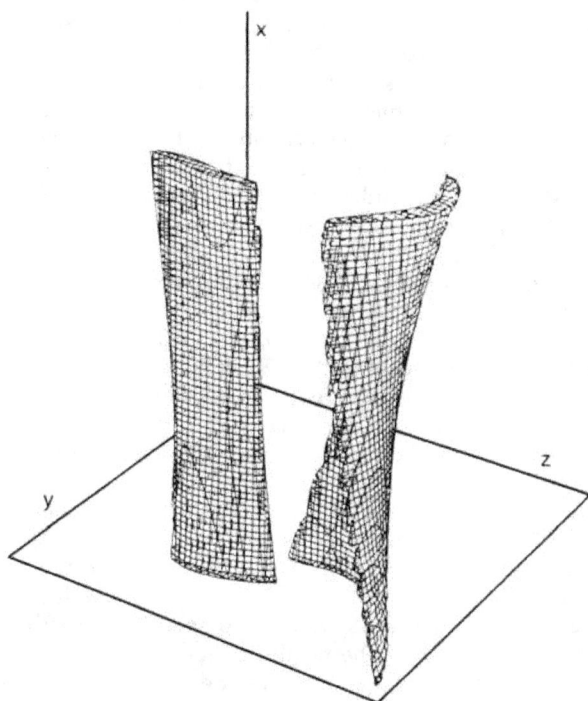

Figure 8-6. Surfaces of constant parallel current density after several steps in the sequence given by Eqs. (8.8) and (8.9).

8.4 An Analogy Between the Solar Corona and the RFP

It is interesting to note that oscillations consisting of cyclical diffusion, instability, and relaxation as seen in the RFP are similar to oscillations that occur in the solar corona [Heyvaerts and Priest, 1984; Berger, 1984; Browning, 1988]. In the corona the instability may be triggered by external driving (photospheric motions) rather than diffusion, and magnetic energy dissipation during relaxation may provide a source of heating, solar flares, or coronal mass ejections. In the corona, resistive diffusion of the magnetic field is too slow to counterbalance the rate at which magnetic energy is fed into the corona by boundary motions, unless very small scale-lengths ($\approx$ cm) or highly anomalous resistivity are hypothesized.

Taylor's relaxation theory has been applied by several authors to explain processes of magnetic reconnection and energy release in the solar corona [Heyvaerts and Priest, 1984; Browning, 1988; Vekstein, 1990], and the MHD computations described in Section 8.2 have shown the general features of the dynamics of solar corona magnetic arcades [Mikić et al., 1988; Biskamp and Welter, 1989]. Like the RFP, the solar corona can be viewed as a driven system; the energy input is supplied by photospheric motions.

However, a significant difference between relaxation in the solar corona and the RFP is the type of drive, or power input. In the solar corona the power input is greater than what is required to sustain the system against resistive diffusion. It is this excess power that drives the system away from the relaxed state; the coronal resistivity is too small for resistive diffusion to have significant global effects. The resulting relaxation may manifest itself as a solar flare or coronal mass ejection. In the RFP the drive is the power supplied by the toroidal voltage that is required to maintain a steady mean field profile. This power input is just sufficient to generate the required level of MHD fluctuations necessary for the dynamo; the system would not be sustained without this power input. The excursions away from the relaxed state are caused here by resistive diffusion, as discussed in Section 4.3.

A steady-state is achieved in the corona by means of reconnection processes that keep pace with the wrapping and twisting of field lines produced by convective motions in the photosphere. In the RFP, reconnection is the result of the MHD fluctuations that maintain the mean field profiles for time periods extending beyond the resistive diffusion time.

CHAPTER 9

SUMMARY

Relaxation, or self-organization, is a process by which a continuous system reaches some preferred, non-trivial, global configuration subject to certain constraints. Mathematically, this is posed as a problem in the calculus of variations in which one integral is minimized subject to the constraint that another integral remain constant. For the case of a magnetized plasma, a useful procedure is to minimize the magnetic energy while the total magnetic helicity is held fixed. The resulting Euler equation is a partial differential equation for the magnetic field whose solutions describe the preferred equilibrium states of the system. These solutions can be characterized by a handful of parameters, such as total magnetic flux and electric current, that may be externally controlled and measured. The locus of these solutions in this low dimensional parameter space defines the operating regimes of many present laboratory plasma experiments.

The relaxed states described above are attainable only in the time-asymptotic sense: they are observed to result from certain dynamical processes that act over a finite time interval. The validity of the variational theory requires only that the underlying dynamics be describable by resistive magnetohydrodynamics. The theory does not address the question of which modes are involved in the relaxation process.

In this book we have attempted to demonstrate how plasma relaxation, or self-organization, can take place as the result of low frequency, long wavelength MHD modes. This point of view is contrary to that taken in fluid mechanics where relaxed states are assumed to arise from small scale turbulence. Early concepts of the cause of relaxation in magnetized plasmas also invoked turbulent fluctuations in a resistive medium; the original relaxation theory was predicated on such a mechanism [Taylor, 1974].

The picture presented here has arisen as a result of extensive study over more than a decade by several research groups, and agrees at least qualitatively with experimental observations in Reversed-field Pinch and other plasmas. This picture has been obtained primarily through experimental measurements and the application of advanced methods of large scale numerical simulation. At this point in time it is unclear whether any lack of quantitative agreement between theory and experiment is due to fundamental physics that is missing from the mathematical model (i.e., resistive magnetohydrodynamics), or merely from the inability of modern computers to accurately model turbulence at large Lundquist numbers.   A resolution of this question awaits advances in computing technology.

Throughout, we have used the Reversed-field Pinch as a paradigm for understanding plasma relaxation. This is because the dynamics of relaxation have been most extensively studied in this device. Similar dynamics seem to occur in the solar corona. Of course, we cannot rule out the possibility that other dynamical models may be important in other circumstances. Perhaps the insights gained in studies of the RFP will prove useful when other configurations are studied in detail.

9.1 Relaxation in the Reversed-field Pinch

The magnetic field configuration in the Reversed-field Pinch (RFP) is sustained much longer than is possible in the presence of resistive diffusion; a dynamo-like process is at work in the pinch. The field is observed to be generated in a sequence of oscillations that, under some circumstances, have a sawtooth-like character; plasma relaxation (the so-called dynamo) is associated with these fluctuations. In Chapter 4 we presented a phenomenological picture of this relaxation in the RFP. This model was formulated strictly on the basis of experimental observations and linear stability theory, and in some cases predated the results of the nonlinear MHD simulations that were presented in Chapter 5. The accuracy of the phenomenological model is well borne out by those calculations.

It is observed that the dominant magnetic fluctuations in the RFP are characterized by poloidal mode number $m = 1$, and toroidal mode number $n \approx -3/2\varepsilon$, where $\varepsilon = a/R$ is the inverse aspect ratio of the torus; they are long wavelength and low frequency. Furthermore, bursts of these fluctuations are associated with relaxation. By fitting parametric models to experimental data, it was determined that during these relaxation events, the amount of current density in the core of the discharge is limited to a maximum value corresponding to $q(0) \gtrsim 2\varepsilon/3$. Linear stability analysis of the model profiles determined that when this limit is exceeded, the configuration becomes linearly unstable to long wavelength resistive MHD modes of the type just mentioned. Furthermore, studies of resistive diffusion of these profiles revealed that this naturally occurring process could drive the profiles unstable on a time scale consistent with the observed sawtooth oscillations. Presumably, in their nonlinear state these unstable modes conspire to generate the magnetic flux necessary to restore the preferred RFP profiles. On the basis of these results, a cyclical model describing the interaction between diffusion and instability was proposed.

The predictions of the phenomenological model are in substantial agreement with the picture obtained from large scale numerical simulation. This was described in detail in Chapter 5. The RFP dynamo, and the ensuing relaxation, can be self-consistently described by a nonlinear resistive MHD model. The dynamo operates because of plasma instability. Within the discharge, there is a constant conflict between the forces of transport attempting to destroy the pinch, and the forces of instability attempting to restore the pinch to a more favored state. The RFP owes its existence to a stand-off between these competing processes.

The fundamental RFP dynamo mechanism is as follows. The energy to drive the dynamo comes from the externally applied voltage. The pinch becomes unstable to global $m = 1$ modes. The nonlinear interaction of these modes drives MHD fluctuations that result in an anomalous parallel electric field arising from the mean inductive part of Ohm's law. The phasing between the velocity and magnetic fluctuations is such that this field suppresses parallel current on the axis and drives parallel current at the edge. Thus both current peaking on axis and loss of field reversal at the edge are avoided, and RFP field and current profiles are maintained.

More specifically, global $m=1$ kink modes are driven unstable by current peaking due to a thermal instability that *lowers* $q(0)$. If the principle unstable mode is resonant, its nonlinear evolution proceeds in two steps: a *first reconnection* that removes the original resonance and further lowers $q(0)$; and, a *second reconnection* that restores the original resonance, *raises* $q(0)$, and restores a stable profile. The second reconnection is a driven, nonlinear process. If the principal unstable mode is nonresonant from above, only the second reconnection occurs. In an idealized symmetry a perfect balance can arise, and the pinch can exit as a steady helical Ohmic state: a resistive steady state with flow.

The quasilinear evolution of $m = 1$ resistive kink modes is sufficient to produce the RFP dynamo, and to produce a helical Ohmic state. Nonlinear mode coupling to other $m = 1$ modes through the driven $m = 0$ modes enhances the dynamo and produces the dynamical details that are observed experimentally. The RFP is basically a driven, nonlinear, dynamical system. The role of linear theory is to identify the important modes. From these results, there is no evidence that small scale turbulence plays a fundamental role in the relaxation process in the RFP.

In addition to describing the basic dynamical mechanism for sustainment, the resistive MHD model can also self-consistently describe details of the spectra of magnetic fluctuations, the origin of apparently anomalous plasma resistance, voltage and fluctuation increases in the presence of resistive walls and limiters, sawtooth oscillations, slinky modes, and can make predictions for DC helicity injection. Although not discussed here, the MHD model is also able to explain the details of the relaxation process in Spheromaks [*Sgro et al.*, 1987; *Ono and Katsurai*, 1991].

9.2 Relaxation and Transport

Plasma relaxation requires MHD fluctuations. In the RFP, these are the dominant fluctuations that are observed, and are responsible for the sustainment of the discharge: the plasma could not exist without this persistent churning. It is also known that fluctuations of this sort can cause significant energy transport. It is therefore of some interest to speculate as to the role played by the relaxation process in determining the energy balance in specific plasma configurations. As stated by Sykes and Wesson [*Sykes and Wesson*, 1977], it is not a question of whether the desired magnetic field configuration can be maintained, but rather if the resulting level of anomalous transport is acceptable.

In Chapter 7 it was shown that the MHD modes that are responsible for the dynamo result in a magnetic field that is stochastic over much of the discharge: the magnetic field lines are volume filling rather than surface filling. This type of configuration is known to result in rapid radial energy transport due to the efficiency of heat conduction parallel to the magnetic field. Under some circumstances, a thin region of good flux surfaces (surface filling field lines) may continue to exist near the plasma edge. Such a configuration would be characterized by a broad, flat temperature profile. Any additional magnetic fluctuation that perturbs the edge region may render the discharge completely stochastic. This picture of edge confinement is consistent with the sensitivity of RFP discharges to externally imposed magnetic field perturbations.

Self-consistent numerical simulations of start-up and sustainment of an RFP discharge including thermal evolution, Ohmic heating, and anisotropic heat conduction were presented in Chapter 7. These results exhibited sawtooth oscillations similar to those observed experimentally, and resulted in average values of plasma current, temperature, and energy confinement time that are consistent with experimental results. This is at least suggestive that dynamo fluctuations may play a significant role in determining the thermal transport and confinement properties of the RFP. However, considerably more study is required to make this statement conclusive.

In the quiet solar corona, thin current filaments are naturally generated from long spatial scale boundary motions. The dissipation of these current sheets may result in an anomalously large Ohmic heating power that may contribute to coronal heating. Similarly, in the RFP we may expect, although we have not shown, that the long wavelength dynamo motions may generate current filaments at short spatial scale lengths. Their dissipation may be a significant heating source for hot plasma discharges.

9.3 Relaxation in the Solar Corona

An analogy can be drawn between the dynamics of the RFP and the solar corona, or outer atmosphere of the sun. Both are driven systems that may exhibit cyclical behavior. During part of the cycle the system is driven away from its preferred, relaxed state. In the RFP this is accomplished by resistive diffusion. In the corona the system is driven by convective motions in the photosphere. In the RFP, the preferred state is restored as a result of MHD instabilities that produce the dynamo. In active regions of the corona, the relaxation is driven by MHD instabilities that may produce a solar flare. In both systems the cycle may repeat several times.

In Chapter 8 these ideas were made more concrete by applying the same techniques of large scale numerical simulation that were successful in studies of the RFP to the solar corona. For the model problem of a periodic arcade structure driven by sheared photospheric flow it is possible to slowly store and then rapidly release an amount of magnetic energy that is consistent with a large solar flare. The nonlinear evolution of this structure had many features in common with observations.

Studies of the quiet corona driven by random footpoint motions were also described. These studies are relevant to the topic of coronal heating: the coronal plasma is too conductive to receive substantial heating from Ohmic dissipation of large scale electric currents. Long wavelength motions in the photosphere can naturally generate very thin, filamentary current structures in the corona. The rapid dissipation of these thin current sheets may produce an anomalously large Ohmic heating power that may be at least partially responsible for heating the corona. In a steady state, the rate of work done on the magnetic field by the motion of the photosphere is just balanced by the Ohmic dissipation. Here, relaxation and external driving are occurring on the same time scale.

9.4 Critique

The picture of plasma relaxation presented in this book is suggestive, and perhaps even convincing, but it is not absolutely conclusive. It seems clear that, at least in the RFP, there is good agreement between theory and experiment with regard to the dominant magnetic fluctuations. In numerical simulations, these fluctuations can generate a dynamo and self-consistently reproduce many of the experimental properties of RFP discharges. However, some questions remain. These are discussed below.

To what extent can the numerical simulations be believed?

The results presented in this book are based largely upon three-dimensional, time-dependent, nonlinear, long time scale, numerical solution of the resistive MHD equations. Such calculations are extremely difficult. The codes that are used are quite complicated, and have taken many man-years to develop. They have been benchmarked against known solutions, and against each other [Aydemir et al., 1985; Strauss, 1985]. *The results of these calculations agree well with those expected on the basis of phenomenological models, and with experiment.* However, one should always view the results of any numerical calculation with caution. They should be judged in the overall context defined by experiment and analytic theory.

What about analytic theory?

One would feel more comfortable with the numerical results if they were supported by an analytical theory. However, the simulations have suggested that relaxation, at least in the RFP, is inherently nonlinear, and includes the interaction of many modes. This fact makes the development of an analytic theory extremely difficult. To date, no satisfactory nonlinear analytic theory of the RFP dynamo has been given. There is a large gap between the simple variational description of Taylor and the detailed, three-dimensional numerical calculations. Perhaps this book will serve as a motivation for the development of such a theoretical framework.

What is the role of turbulence?

The effect of small scale turbulence cannot be directly assessed from the present results because of insufficient spatial resolution in the numerical simulations. Fully turbulent numerical MHD simulations of naturally occurring and laboratory plasmas are beyond the capability of present computing technology. Thus, relaxation as the result of small scale turbulence alone cannot be completely ruled out. However, it has been demonstrated in this book that long wavelength motions can in fact produce plasma relaxation that is very similar to that observed experimentally. It would be very surprising if this result changed fundamentally as more modes are added to the simulations. We remark that attempts to simulate turbulent relaxation in the RFP [*Dahlberg et al.*, 1986] have resulted in the generation of global helical structures that are topologically similar to the long wavelength dynamo modes reported in the previous chapters. Whether this is interpreted as a self-organization of the turbulence or the onset of a global $m = 1$ instability is likely moot.

Are pressure driven modes important?

The results presented in this book have been derived for the most part from the force-free resistive MHD model described in Chapter 2. This precludes the appearance of pressure driven modes, and it is natural to inquire what role they may play in plasma relaxation. In Chapter 7 we described relaxation calculations with finite plasma pressure. Benchmark tests of the code employed for these calculations clearly indicated that unstable pressure driven modes could be identified and calculated. However, during the relaxation simulations only the current driven kink modes appeared. This may be the result of the flat pressure profile in the core of the plasma that results from the stochastic magnetic field. In any case, the pressure driven modes, if they exist, are relegated to the plasma edge where they may play a role in thermal transport; they do not appear to be significant in determining the major features of the dynamics of plasma relaxation.

Are there non-MHD effects?

Clearly, physical effects occur in hot plasmas that are not describable by resistive MHD. Theories of the dynamo based on kinetic effects that do not appear in the resistive MHD equations have been formulated [*Jacobsen and Moses*, 1984]. These theories are known as the *kinetic dynamo*. They depend on the existence of a non-local Ohm's law that arises because of the stochastic fields in the core of the plasma. However, dynamo theories based on these effects are not self-consistent; they require the postulation of physical effects, such as magnetic field fluctuations and stochastic field lines, that are outside the model. As a result, they cannot account for the variety of experimental observations that are predicted self-consistently by the resistive MHD theory.

What is the future of relaxation studies?

The magnetohydrodynamic theory of plasma relaxation has become quite mature. However, as pointed out in the previous paragraphs, the picture is still not complete. For example, a dynamical analytic theory based on insights gained from the numerical simulations would be very useful.

This sort of synergism between simulation, experiment, and theory is required to make progress on difficult, nonlinear problems.

An important practical issue, at least in the RFP, is to quantify the role of the dynamo fluctuations in transport and heating. This requires further detailed experimental studies of hot plasmas and extensive application of already existing computing technology to these problems.

Finally, we have concentrated primarily on relaxation in the RFP. It may be that nature will choose other avenues to the relaxed state in other magnetoplasma configurations. These must be explored with the same intensity as was required to produce the successful model of MHD relaxation that has been presented here.

References

Alper, B., M. K. Bevir, H. A. B. Bodin, C. A. Bunting, P. G. Carolan *et al.* (1989a) in *Plasma Physics and Controlled Nuclear Fusion Research 1988*, (IAEA, Vienna) Volume 2, p. 431.

Alper, B., M. K. Bevir, H. A. B. Bodin, C. A. Bunting, P. G. Carolan *et al.* (1989b) *Plasma Phys. and Cont. Fusion* **31**, 205.

Alper, B., and P. Martin (1989) in *Proc. 16th Europ. Conf. on Cont. Fusion and Plasma Physics, Venice 1989*, EPS, Volume 2, p. 725.

Antiochos, S. K. (1987) *Ap. J.* **312**, 886.

Antoni, V., D. Merlin, S. Ortolani, and R. Paccagnella (1986) *Nucl. Fusion* **26**, 1711.

Antoni, V., and S. Ortolani (1987) *Phys. Fluids* **30**, 1489.

Antoni, V., P. Martin, and S. Ortolani (1989) *Nuc. Fusion* **29**, 1759.

Asakura, N., A. Fujisawa, T. Fujita, Y. Fukuda, H. Hattori *et al.* (1987) in *Plasma Physics and Controlled Nuclear Fusion Research, 1986* (IAEA, Vienna) Volume 2, p. 433.

Aydemir, A. Y., and Barnes, D. C. (1984) *Phys. Rev. Lett.* **52**, 930.

Aydemir, A. Y., D. C. Barnes, E. J. Caramana, A. A. Mirin, R. A. Nebel, D. D. Schnack, and A. G. Sgro (1985) *Phys. Fluids* **28**, 898.

Batacharjee, A., R. L. Dewar, and D. Monticello (1980) *Phys. Rev. Lett.* **45**, 347.

Battacharjee, A., and R. L. Dewar (1982) *Phys. Fluids* **25**, 887.

Bateman, G. (1978) *MHD Instabilities* (MIT Press, Cambridge).

Berger, M. A., (1984) *Geophys. Astrophys. Fluid Dynamics* **30**, 63.

Berger, M. A., and G. B. Field (1984) *J. Fluid Mech.* **147**, 133.

Bernstein, I. B., E. A. Freiman, M. D. Kruskal, and R. M. Kulsrud (1958) Proc. Roy. Soc. London. Ser. A **244**, 17.

Bevir, M., and J. Gray (1980) in *Proc. of the Reversed Field Pinch Theory Workshop*, H. R. Lewis and R. A. Gerwin, Eds., Los Alamos Natl. Lab. Report LA-8944-C (Los Alamos, NM) p. 176.

Bickerton, R. J., F. Alladio, D. V. Bartlett, K. Behringer, R. Behrinson *et al.* (1986) *Plasma Physics and Cont. Fusion* **28**, 55.

Biskamp, D., and Welter, H. (1989) *Solar Phys.* **120**, 49.

Bodin, H. A. B., and A. A. Newton (1980) *Nucl. Fusion* **20**, 1255.

Braginskii, S. I. (1965) in *Reviews of Plasma Physics* (Consultants Bureau, New York) Volume 1.

Browning, P. K. (1988) *Plasma Physics and Controlled Fusion* **30**, 1.

Burton, W. M., E. P. Butt, H. C. Cole, A. Gibson, D. W. Mason, and R. S. Pease (1962) *Nucl. Fusion Supp.* Part 3, 903.

Butt , E. P., R. Carruthers, J. T. D. Mitchell, R. S. Pease, P. C. Thonemann *et al.*, (1958) in *Peaceful Uses of Atomic Energy* (Proc. 2nd Int. Conf., UN, Geneva, 1958) **32**, 42.

Caramana, E. J., R. A. Nebel, and D. D. Schnack (1983) *Phys. Fluids* **26**, 1305.

Caramana, E. J., and D. A. Baker (1984) *Nucl. Fusion* **24**, 423.

Caramana, E. J., and D. D. Schnack (1986) *Phys. Fluids* **29**, 3023.

Caramana, E. J. (1989) *Phys. Fluids B* **1**, 2186.

Colgate, S. A., J. F. Ferguson, and H. P. Furth (1958) in *Peaceful Uses of Atomic Energy* (Proc. 2nd Int. Conf., UN, Geneva, 1958) **32**, 129.

Cowling, T. G. (1934) Monthly Notices Roy. Astron. Soc. **94**, 39.

Cowling, T. G. (1957) *Magnetohydrodynamics* (Interscience Publishers, New York).

Craddock, G. G. (1991) *Phys. Fluids B* **3**, 316.

Cunnane, J. A., D. E. Evans, C. G. Gimblett, T. C. Hender, and H. Y. W. Tsui (1988) in *Physics of Mirrors, Reversed Field Pinches, and Compact Tori* (Proc. Int. School of Plasma Physics, Varenna, 1987) Volume 3, p. 1017.

Dahlburg, J. P., D. Montgomery, G. D. Doolen, and L. Turner (1986) *Phys. Rev. Lett.* **57**, 428.

Dobrott, D. D., D. C. Barnes, Z. Mikić, and D. D. Schnack (1985) *Bull. Am. Phys. Soc.* **30**, 1399.

Drake, J. F., P. L. Pritchett, and Y. C. Lee (1978) "Nonlinear Evolution of Tearing Instabilities: Violations of Constant Psi", UCLA Report PPG-341 (unpublished).

Fernandez, J. C., B. L. Wright, G. J. Marklin, D. A. Platts, and T. R. Jarboe (1989) *Phys. Fluids B* **1**, 1254.

Finn, J. M., R. A. Nebel, and C. Bathke (1992) *Phys. Fluids B* **4**, 1262.

Fornberg, B. (1977) *J. Comp. Phys.* **25**, 1.

Freidberg, J. P. (1987) *Ideal Magnetohydrodynamics* (Plenum Press, New York).

Frisch, U., A. Pouquet, J. Leorat, and A. Mazure (1975) *J. Fluid Mech.* **68**, 769.

Furth, H., J. Killeen, and M. N. Rosenbluth (1963) *Phys. Fluids* **6**, 459.

Furth, H. P. (1969) in *Plasma Instabilities in Astrophysics*, D. G. Wentzel and D. A. Tidman, Eds. (Gordon and Breach, New York).

Furth, H. P. (1985) *Phys. Fluids* **28**, 1595.

Gibson, R. D., and K. Whiteman (1968) *Plasma Phys.* **10**, 1101.

Gimblett, C. G., and M. L. Watkins (1975) *Cont. Fusion and Plasma Physics* **1**, 103.

Gimblett, C. G. (1980) in *Proc. of the Reversed Field Pinch Theory Workshop*, H. R. Lewis and R. A. Gerwin, Eds., Los Alamos Natl. Lab. Report LA-8944-C (Los Alamos, NM) p. 254.

Gimblett, C. G. (1986) *Nucl. Fusion* **26**, 617.

Gimblett, C. G., P. J. Hall, J. B. Taylor, and M. F. Turner (1987) *Phys. Fluids* **30**, 3186.

Hagyard, M. J., J. B. Smith, Jr., D. Teuber, and E. A. West (1984) *Solar Phys.* 91, 115.

Hameiri, E., and J. H. Hammer (1982) *Phys. Fluids* 25, 1855.

Harned, D. H., and W. Kerner (1985) *J. Comp. Phys.* 60, 62.

Harned, D. H., and W. Kerner (1986) *Nucl. Sci. and Engrg.* 92, 119.

Harned, D. H., and D. D. Schnack (1986) *J. Comp. Phys.* 65, 57.

Hasegawa, A. (1985) *Advances in Physics* 34,1.

Heyvaerts, J., and E. R. Priest (1984) *Astron. Astrophys.* 137, 63.

Hirano, Y., Y. Kondoh, Y. Maejima, Y. Nogi, K. Ogawa *et al.* (1985) in *Plasma Physics and Controlled Nuclear Fusion Research 1984* (IAEA, Vienna) Volume 2, p. 475.

Ho, Y. L., and S. C. Prager (1988) *Phys. Fluids* 31, 1673.

Ho, Y. L., S. C. Prager, and D. D. Schnack (1989) *Phys. Rev. Lett.* 62, 1504.

Ho, Y. L. (1991) *Nucl. Fusion* 31, 341.

Ho, Y. L., and G. G. Craddock (1991) *Phys. Fluids* B 3, 721.

Ho, Y. L., and S. C. Prager (1991) *Phys. Fluids* B 3, 3099.

Hokin, S., A. Almagri, S. Asadi, J. Beckstead, G. Chartas *et al.* (1991) *Phys. Fluids* 3, 2241.

Holmes, J. A., B. A. Carraras, P. H. Diamond, and V. E. Lynch (1988) *Phys. Fluids* 31, 1166.

Hossain, M., W. H. Matthaeus, and D. Montgomery (1983) *J. Plasma Phys.* 30, 479.

Hutchinson, I. H., M. Malacarne, P. Noonan, and D. Brotherton-Radcliffe (1984) *Nucl. Fusion* 24, 59.

Jacobson, A. R., and R. W. Moses (1984) *Phys. Rev. A* 29, 3335.

Jarboe, T. R., I. Henins, A. R. Sherwood, C. W. Barnes, and H. W. Hoida (1983) *Phys. Rev. Lett.* 51, 39.

Kadomtsev, B. B. (1975) *Sov. J. Plasma Phys.* 1, 389.

Kadomtsev, B. B. (1977) in *Plasma Physics and Controlled Nuclear Fusion Research 1976* (IAEA, Vienna) Volume 1, p 555.

Kadomtsev, B. B. (1987) in *Proc. 1987 Int. Conf. on Plasma Physics*, A. G. Sitenko, Ed. (World Scientific Press, Singapore) Volume 2, p. 1273.

Kailhacker, M., *et al* (1985) in *Plasma Physics and Controlled Nuclear Fusion Research 1984* (IAEA, Vienna) Volume 2, p. 69.

Königl, A., and A. R. Chouduri (1985) *Ap. J.* 289, 173.

Kraichnan, R. H. (1967) *Phys. Fluids* 10, 1417.

Kraichnan, R. H. (1973) *J. Fluid Mech.* 59, 745.

Krall, N. A., and A. W. Trivelpiece (1973) *Principles of Plasma Physics* (McGraw-Hill, New York).

Krause, F., and H. K. Rädler (1980) *Mean Field Electrodynamics and Dynamo Theory* (Pergamon Press, New York).

Kusano, K., and T. Sato (1987) *Nucl. Fusion* 27, 821.

Kusano, K., T. Tamano, and T. Sato (1991) *Nucl. Fusion* **31**, 1923.

LaHaye, R. J., T. N. Carlson, R. R. Goforth, G. L. Jackson, M. J. Schaffer, T. Tamano, and P. L. Taylor (1984) *Phys. Fluids* **27**, 2576.

LaHaye, R. J., T. H. Jensen, P. S. C. Lee, R. W. Moore, and T. Ohkawa (1986) *Nucl. Fusion* **26**, 255.

Landau, L. D., and E. M. Lifshitz (1959) *Fluid Mechanics* (Pregamon Press, Oxford).

Malesani, G. (1988) in *Physics of Mirrors, Reversed Field Pinches, and Compact Tori* (Proc. Int. School of Plasma Physics, Varenna, 1987) Volume 1, p. 331.

Martin, T. J., and J. B. Taylor (1974) "Helically Deformed States of Toroidal Pinches", Culham Laboratory Report (unpublished).

Massey, R. S., R. G. Watt, P. G. Weber, G. A. Wurden, D. A. Baker *et al.* (1985) *Fusion Technol.* **8**, 1571.

Mikić, Z., D. C. Barnes, and D. D. Schnack (1988) *Ap. J.* **328**, 830.

Mikić, Z., D. D. Schnack, and G. Van Hoven (1989) *Ap. J.* **338**, 1148.

Moffatt, H. K. (1969) *J. Fluid Mech.* **35**, 117.

Moffatt, H. K. (1978) *Magnetic Field Generation in Electrically Conducting Fluids* (Cambridge University Press, Cambridge).

Montgomery, D., L. Turner, and G. Vahala (1978) *Phys. Fluids* **21**, 757.

Montgomery, D., and L. Phillips (1988) *Phys. Rev. A* **38**, 2953.

Montgomery, D. (1989) "Relaxed States of Driven, Dissipative Magnetohydrodynamics: Helical Distortions and Vortex Pairs", presented at Univ. of Minnesota Colloquium *Trends in Theoretical Physics*, May 18, 1989 (unpublished).

Murakami, M., V. Arunasalam, J. D. Bell, M. G. Bell, M. Bitter *et al.* (1985) *Plasma Physics and Cont. Fusion* **28**, 17.

Nebel, R. A., E. J. Caramana, and D. D. Schnack (1989) *Phys. Fluids B* **8**, 1671.

Neidig, D. F., DeMastus, H. L., and Wiborg, H. P. (1978) Air Force Geophysics Laboratory Report AFGL-TR-1094, Hanscom AFB, MA.

Newcomb, W. A. (1960) *Ann. Phys.* **10**, 232.

Ohkawa, T., H. K. Forsen, A. A. Schupp, and D. W. Kerst (1963) *Phys. Fluids* **6**, 846.

Ono, Y., and M. Katsurai (1991) *Nucl. Fusion* **31**, 233.

Ortolani, S. (1984) in *Course on Mirror-Based and Field-Reversed Approaches to Magnetic Fusion* (Proc. Int. School of Plasma Physics, Varenna, 1983) Volume 2, p. 513.

Ortolani, S. (1985) in *Twenty Years of Plasma Physics*, B. McNamara, Ed., (World Scientific Press, Singapore) p.75.

Ortolani, S. (1987) in *Plasma Physics* (Proc. 7th Int. Conf. Kiev, 1987), (World Scientific Press, Singapore) p. 802.

Parker, E. N. (1972) *Ap. J.* **174**, 499.

Parker, E. N. (1983) *Ap. J.* **264**, 635.

Parker, E. N. (1986) *Geophys. Astrophys. Fluid Dyn.* **35**, 243.

Phillips, J. A., D. A. Baker, R. F. Gribble, and C. Munson (1988) *Nucl. Fusion* **28**, 1241.

Pouquet, A., U. Frisch, and J. Leorat (1976) *J. Fluid Mech.* **77**, 321.

Prager, S. C. (1990) private communication.

Priest, E. R. (1982) *Solar Magnetohydrodynamics* (Reidel, Dordrecht)

Rechester, A. B., and M. N. Rosenbluth (1978) *Phys. Rev. Lett* **40**, 38.

Reiman, A., (1980) *Phys. Fluids* **23**, 230.

Robinson, D. C. (1971) *Plasma Physics* **13**, 439.

Robinson, D. C. (1978) *Nucl. Fusion* **18**, 939.

Rosenbluth, M. N. (1958) in *Peaceful Uses of Atomic Energy* (Proc. 2nd Int. Conf., UN, Geneva, 1958) **31**, 42.

Rusbridge, M. G., D. J. Lees, and P. A. H. Saunders (1962) *Nucl. Fusion Supp.* **3**, 895.

Rutherford, P. H. (1973) *Phys. Fluids* **16**, 1903.

Schnack, D. D., D. C. Baxter, and E. J. Caramana (1984) *J. Comp. Phys.* **55**, 485.

Schnack, D. D., E. J. Caramana, and R. A. Nebel (1985) *Phys. Fluids* **28**, 321.

Schnack, D. D., D. C. Barnes, Z. Mikić, D. S. Harned, E. J. Caramana, and R. A. Nebel (1986) *Comp. Phys. Comm.* **43**, 17.

Schnack, D. D., D. C. Barnes, Z. Mikić, D. S. Harned, and E. J. Caramana (1987) *J. Comp. Phys.* **70**, 330.

Schnack, D. D., and S. Ortolani (1990) *Nucl. Fusion* **30**, 277.

Schnack, D. D. (1991) in *Physics of Alternative Magnetic Confinement Schemes* (Proc. Int. School of Plasma Physics, Varenna, 1991) p. 631.

Scime, E., S. Hokin, N. Mattor, and C. Watts (1992) *Phys. Rev. Lett.* **68**, 2165.

Sgro, A. G., A. A. Mirin, and G. Marklin (1987) *Phys. Fluids* **30**, 3219.

Strauss, H. R. (1985) *Phys. Fluids* **28**, 2786.

Sturrock, P. A. (1980) in *Solar Flares*, P. A. Sturrock, Ed. (Associated University Press, Boulder) p. 411.

Sykes, A., and J. A. Wesson (1977) in *Proc. 8th European Conf. on Cont. Fusion and Plasma Physics*, Prague, (Czechoslavak Acadamy of Sciences, 1977) p. 80.

Tamano, T., W. D. Bard, C. Chu, Y. Kondoh, R. J. LaHaye, P. S. Lee, M. T. Saito, M. J. Schaffer, and P. L. Taylor (1987) *Phys. Rev. Lett.* **59**, 1444.

Taylor, J. B. (1974) *Phys. Rev. Lett.* **33**, 139.

Taylor, J. B. (1975) in *Plasma Physics and Controlled Nuclear Fusion Research 1974* (IAEA, Vienna) Volume 1, p. 161.

Taylor, J. B. (1980) in *Proc. of the Reversed Field Pinch Theory Workshop*, H. R. Lewis and R. A. Gerwin, Eds., Los Alamos Natl. Lab. Report LA-8944-C (Los Alamos, NM) p. 239.

Taylor, J. B. (1984) in *Course on Mirror-Based and Field-Reversed Approaches to Magnetic Fusion* (Proc. Int. School of Plasma Physics, Varenna, 1983) Volume 2, p. 501.

Taylor, J. B. (1986) *Rev. Mod. Phys.* **58**, 741.

Taylor, J. B., and M. F. Turner (1989) *Nucl. Fusion* **29**, 219.

Tsui, H. Y. W., and J. A. Cunnane (1988) *Plasma Physics and Controlled Fusion* **30**, 865.

Ueda, Y., N. Asakura, S. Matsuzuka, K. Yamagishi, S. Shinohara, Y. Nagayama, H. Toyama, K. Miyamoto, and N. Inoue (1987) *Nucl. Fusion* **27**, 1453.

van Ballegooijen, A. A. (1985) *Ap. J.* **298**, 421.

van Ballegooijen, A. A. (1986) *Ap. J.* **311**, 1001.

van Ballegooijen, A. A. (1988a) *Geophys. Astrophys. Fluid Dyn.* **41**, 181.

van Ballegooijen, A. A. (1988b) in *Proc. 9th Sacramento Peak Summer Workshop on Solar and Stellar Coronal Structure and Dynamics*, R. Altrock, Ed. (Nat. Solar Observatory, Sunspot, NM) p. 115.

Van Hoven, G., S. S. Ma, and G. Einaudi (1981) *Astron. Astrophys.* **97**, 232.

Vekstein, G. E. (1990) *Astron. Astrophys.* **182**, 405.

Voslamber, D., and D. K. Callabaut (1962) *Phys. Rev.* **128**, 2016.

Watt, R. G., and R. A. Nebel (1983) *Phys. Fluids* **26**, 1168.

Werley, K. A., R. A. Nebel, and G. A. Wurden (1985) *Phys. Fluids* **28**, 1450.

Wesson, J. A. (1979) private communication.

Woltjer, L. (1958) *Proc. Natl. Acad. Sci. USA* **44**, 489.

Wurden, G. A. (1984) *Phys. Fluids* **27**, 551.

Zweibel, E. G. and Li, H. S. (1987) *Ap. J.* **312**, 423.

Index

A

Alpha-Effect,　102, 110
Active Regions,　155, 157–158, 175
Adiabatic Law,　21
Alfvén Waves,　24, 37, 44, 53
Alfvén Transit Time,　25, 31, 104–107
Arcade, 156–163, 175
Aspect Ratio,　10–11, 39, 44, 89, 104,
　　114–118, 122, 125, 151, 172
Astrophysical Dynamo,　96, 103

B

Beta-Effect,　102
Bessel Function,　4, 59, 68, 104, 112, 132
Bessel Function Model (BFM),
　　4, 59–61, 68, 70–76, 80–83
Beta,　10–11, 151

C

Cascade,　6–7, 26, 38, 164–166
Classical Dynamo Theory,　96, 102–103
Compressional Alfvén Waves,　24
Constant-ψ,　38
Constraints,　1, 6, 26–27, 48–53, 68, 74, 171
Continuity Equation,　17
Coronal Dynamics,　155, 163
Coronal Heating,　155, 157, 163–164, 175
Coulomb Collisions,　130, 144, 154
Cowling's Theorem,　99–102
Critical Profile,　8
Current Driven Modes,　28, 40, 79–80, 85
Current Filaments,　163, 166, 174–175
Cyclical Model,　87, 172

D

Disruption,　6, 27, 37, 163
Dominant Dynamo Mode,　114
Driven Reconnection,　106–109
Dynamo,　12, 30, 40, 96–103, 115, 118,
　　120–122, 129–130, 133–136, 139–140,
　　145–148, 152, 170–178

E

Electromagnetic Fields,　21

Energy Confinement Time,
　　144, 148–151, 174
Energy Equation,　21, 43, 58, 144, 149
Energy Principle,　40
Enstrophy,　7
Entropy,　1, 64
Equation of Motion,
　　17–19, 22, 97, 103, 149
Equation of State,　20, 43
Equilibrium,　11, 16, 27, 32, 42, 47, 52, 55,
　　58, 64–65, 69, 75, 84, 87, 95, 100,
　　113–115, 145, 164–166, 171
Ergodically,　30, 49, 145
Euler Equation,　41, 171
Eulerian,　18–20
Experimental Observations,
　　12, 67–68, 88, 92, 95, 126, 136, 144,
　　150, 171–172, 177
Experimental Results,　5, 9, 12, 67, 70, 74,
　　88, 95, 113, 136, 145, 152, 174
External Modes,　76, 82

F

F,　29, 64, 112, 125–126
Fluctuations,　1, 6, 8, 13, 15, 27, 44, 47, 56,
　　64, 88–89, 92, 95, 102–103, 106,
　　110–111, 115–118, 127, 129–130,
　　133–140, 143–145, 148, 152, 154–155,
　　171–175, 178
Flux Surfaces,　10, 28, 30, 108–109, 114, 143,
　　145–150, 174
Flux Tubes,　3, 5, 7, 49–53, 58
Force Free,　3, 5, 11, 42–43, 69, 84–85, 158
Force-Free Fields,　68
Force-Free MHD Model,　44
Free Energy,
　　27–28, 38, 64, 74, 104, 136, 156, 157

H

Helical Ohmic State,
　　113–116, 121–122, 173
Helicity,　6–7, 51, 54, 56–58, 65, 68,
　　107–109, 112, 114, 138–140
Helicity Balance,　127, 129, 138–39
Helicity Injection,　140–141, 174

I

Ideal Magnetohydrodynamic,   2
Ideal MHD,   16, 31, 48, 51–53, 57–58, 68,
        74, 161
Ideal Mode,   75, 80, 82
Incompressible,   19, 24
Incompressible Flow,   34
Interchange Modes,   27–28
Internal Modes,   76, 78, 80, 82, 92, 133
Invariant,
        1– 3, 5–7, 21, 47, 50, 52–57, 60, 111

K

Kinematic Dynamo,   97, 103
Kinematic Dynamo Theory,   97–98
Kinetic Dynamo,   177
Kinetic Energy,   7, 27, 34, 97, 103, 105,
        134– 135, 161, 163
Kink Mode,   11, 30, 104, 106–110, 115–116,
        127, 133, 136, 144, 152, 173, 177
Kruskal-Shafranov Condition,   30

L

Lagrangian,   18–19
Laminar Dynamo,   100
Linear MHD Stability,   67, 87, 130
Linear MHD Stability,   74
Loop Voltage,   13, 100, 103, 127, 129–130,
        133–140, 143
Lundquist Number,   15, 25–26, 36, 44–45, 53,
        121–122, 125, 151, 160

M

Magnetic Arcade,   157–161, 170
Magnetic Energy,   7, 16, 29, 51, 97–98, 103,
        111–112, 121–122, 125–127,
        134–135, 140–141, 156–163, 170–171,
        175
Magnetic Helicity,   3, 7, 47, 54, 56, 57, 59,
        65, 111, 127, 138, 171
Magnetic Island,   32–33, 37–38, 108–109,
        145, 148
Magnetic Reconnection,
        5, 8, 42, 56, 107, 161, 170
Magnetic Shear,   11, 28–30
Magnetoacoustic Waves,   24
Mean Field,   67–68, 74, 82, 87– 88, 101–104,
        110, 118, 120–125, 138–140, 152, 170

MHD Equations,   25–26, 31–32, 35–38, 40,
        43–44, 65, 67, 75, 95, 104, 107, 115,
        120, 164–165, 176–177
MHD Fluctuations,   8, 43, 115–116, 130,
        154, 170, 173–174
Minimization,   48, 51, 54–55, 58
Minimum Energy States,   60, 68
Mode Coupling,   37, 118, 124
Multipinch,   48, 62– 64

N

Navier-Stokes,   1, 7
Non-Constant-ψ,   38, 40
Non-Mhd Effects,   177
Nonlinear Dynamo,   124, 136
Nonlinear Effects,   37, 121
Nonlinear Mode Coupling,
        115–116, 118, 122, 124–125, 137, 173
Normal Modes,   8, 12, 22–23, 26–31, 40,
        43–44, 115
Numerical Simulation,
        8–9, 15, 43–44, 56, 58, 65, 67, 74, 80,
        95–96, 104, 109, 111–112, 115, 120,
        134, 137, 139, 146, 149, 151, 155, 157,
        171, 173–177

O

Ohm's Law,   22, 35, 53, 63, 84, 97, 102–103,
        110, 160, 173, 177

P

Paramagnetic Model,   84–85
Photosphere,   155–158, 160, 163, 165, 170,
        175
Pinch Parameter,   4, 60, 62, 122–123, 151
Plasmoid,   161–163
Pressure,   2, 5, 6, 8, 11, 16–18, 21, 25,
        27– 28, 43, 48, 52, 58, 62–64, 68–70,
        95, 97, 144, 149–153, 160, 164, 177
Pressure Driven Modes,   28, 177
Profile Consistency,   6

Q

Quasilinear Coupling,   118
Quasilinear Effect,   38, 121

R

Rational Surface,   30, 41–42, 75
Reconnection,
        5–6, 12, 33, 42, 53, 57–58, 87, 92,
        107–108, 110, 16–164, 170, 173
Relaxation Mechanism,   8, 104
Relaxation Time,   44–45
Relaxed States,   3–5, 8, 47, 52, 55, 58,
        62–63, 67–68, 74–75, 82, 95, 144, 171
Resistive Diffusion,
        8, 11, 25–26, 31, 36, 44, 57, 67–68,
        82,–83, 85–88, 95–96, 101–103, 106,
        108–112, 125, 127, 129, 138–139, 144,
        151–153, 156, 166, 170, 172, 175
Resistive Diffusion Time,   160
Resistive MHD,
        7, 12, 15, 22, 25–26, 38, 43, 47, 56, 65,
        67, 75, 79, 87, 92, 95–97, 106, 129,
        145, 149, 151, 154, 171, 172–173, 177
Resistive Mode,
        33, 36, 38, 75–76, 80, 82, 92, 137
Resistive Shell,   44, 131–140
Resistive Wall,   95, 127, 130, 146, 173
Resistivity,   3, 6, 12, 16, 21–22, 26, 31–32,
        42, 53, 58, 63, 68, 82–84, 99–100,
        102, 106, 114, 129–131, 144, 149,
        154, 164–165, 170
Resonant Surface,   41, 75
Resonant Surfaces,   30, 145
Reversed-Field Pinch,   1, 58, 88, 129, 144,
        154–156, 171–172
RFP Dynamo,   12, 37, 95–96, 102–103, 108,
        110, 113, 115, 121, 130

S

Safety Factor,   11, 30, 107, 117
Sausage Mode,   30
Sawtooth Oscillations,
        13, 40, 95, 115, 122, 144, 145, 148,
        150–152, 154, 172–174
Self-Organization,   1, 6, 7, 171, 177
Semi-Implicit Methods,   44
Shear Alfvén Waves,   24
Singular Surface,   32–35
Singular Surfaces,   28–29, 74
Slinky Mode,   173
Solar Corona,   13, 16, 155–156, 163–164,
        166, 170, 172, 174–175

Solar Flare,
        16, 37, 156–158, 162, 170, 175
Sound Waves,   24, 26
Spheromak,   1, 174
Stability,   4, 11–12, 15–16, 22, 26, 29, 31,
        36, 37, 39–41, 44, 47, 67, 68–69,
        74–83, 95, 105–106, 121, 124–125,
        133, 165, 172
Stochastic Magnetic Field,
        146–147, 154, 177
Sustainment,   88, 92, 95–96, 103–104,
        111–112, 116–117, 134, 139–140, 143,
        150–154, 173–174

T

Taylor's Conjecture,   53, 55, 58
Taylor's Theory,   2, 8, 12–13, 42, 47, 58,
        60–63, 67–68, 73, 82
Tearing Mode,   32, 36, 37–38, 56, 74, 76, 92
Terrestrial Dynamo,   96, 102
Thermal Conduction,   16, 21, 143–145,
        148–150, 154
Thermal Conductivity,   144, 149
Tokamak,   1, 4–11, 29–30, 39–42, 73–74
Topology,   32, 164
Toroidal Flux,   48, 57, 59–60, 63, 74, 76,
        79–80, 83, 85, 100, 103, 126–127, 130,
        135, 145, 147–148, 151–152
Toroidal Geometry,   58, 64
Transport,   13, 16, 21, 27, 37, 95, 106–107,
        110, 113, 120–121, 129, 140,
        143–145, 14–150, 154, 173–174,
        177–178
Turbulence,   6–9, 45, 56, 95, 101, 103, 171,
        173, 176–177
Turbulent Dynamo,   101–103

V

Variational Principle,   2, 7, 47
Viscosity,   6–9, 16–17, 21, 45, 165
Vorticity,   7

W

Woltjer,   2, 49–50, 68
Woltjer Constraints,   48–50